AGRICULTURAL ENGINEERING

NEW INDIA PUBLISHING AGENCY
New Delhi – 110 034

AGRICULTURAL ENGINEERING
A Practical Approach

P. Kandasamy, Ph.D
Assistant Professor
Department of Agricultural Engineering
Visva-Bharati (A Central University)
Sriniketan-731236, West Bengal

NEW INDIA PUBLISHING AGENCY
New Delhi – 110 034

NEW INDIA PUBLISHING AGENCY
101, Vikas Surya Plaza, CU Block, LSC Market
Pitam Pura, New Delhi - 110 034, India
Email: info@nipabooks.com
Web: www.nipabooks.com

For customer assistance, please contact
Phone: + 91-11-27 34 17 17 Fax: + 91-11- 27 34 16 16
E-mail: feedbacks@nipabooks.com

ISBN : 978-81-94766-80-3

Composed and Designed by NIPA.

Preface

This book Agricultural Engineering: A Practical Approach that integrated post harvest engineering, protected cultivation, land surveying and soil-water conservation engineering. The main aim of this book is to organize the scattered information and provide better practical skill to the students. The book has been mainly designed to serve as a practical reference book for the undergraduate students of Agricultural Science, Horticulture and Agricultural Engineering. Specially, Agriculture Science students will find it very suitable as it includes different topics as per the undergraduate Agricultural Science curriculum. Moreover, the students of Agricultural Polytechnics, Trainees of professional training institutions including post harvest laboratories, KVK, Soil Conservations Training Centres, etc. can also make use of practical reference book.

The book has several practical lessons including moisture content determination, equilibrium moisture content estimation, moisture estimation by moisture meters, determination physical, frictional, thermal and mechanical properties of agricultural materials, drying rate calculations, drying characteristics of grain, evolution of lab model thin layer and fluidized bed drier, performance evaluation of seed cleaner, grader, seed separators, rubber-roll paddy dehusker, maize sheller, groundnut decorticator, paddy parboiling unit, design of bag storage, shallow and deep bins, design of conveying equipment, study of different types of green houses, greenhouse construction, covering materials, design of greenhouse heating and cooling systems, chain survey, cross-staff survey, compass survey, plane table survey, leveling, profile leveling, contouring, irrigation water measurement, design of grassed water ways, design of contour and graded bund, design of bench terrace, estimation of soil loss using Universal Soil Loss Equation.

The author is hopeful that the practical book will serve as a good teaching aid in the field of post harvest engineering, protected cultivation and soil water conservation engineering, and good practical guide for the students. The author is grateful to Prof. Debasish Bhattacharya, Principal, Institute of Agriculture, Visva-Bharati and Dr. Debasis Panda, Department of Crop Physiology for their encouragement in preparation of this work and also thank Er. Aman Kumar, Research Scholar for typing some chapters. The author would welcome suggestions and constructive criticism from the readers to improve the quality of the book.

Dr. P. Kandasamy

Contents

1

Determination of Size and Shape of The Agricultural Materials

Aim

- To determine the size and shape of the biological materials.

Materials required

- Vernier caliper
- Screw gauge
- Food grains

Determination of size of the food materials

Size is an important physical attribute of foods used in screening solids to separate foreign materials, grading of fruits and vegetables, and evaluating the quality of food materials. Various types of cleaning, grading and separation equipment are designed on the basis of size. The size and shape of the grains are used to determine the size of screen openings, angle of inclination, vibration amplitude and frequency of screen in designing an air screen grain cleaner and conveying characteristics of the solid materials.

Measurement of size

Particle sizes are expressed in different units depending on the size range involved. Coarse particles are measured in millimeters, fine particles in terms of screen size, and very fine particles in micrometers or nanometers. The size of materials can be determined using the projected area method. Three characteristic dimensions are:

- **Major diameter:** It is the longest dimension of the maximum projected area
- **Intermediate diameter:** It is the minimum diameter of the maximum projected area or the maximum diameter of the minimum projected area.

- **Minor diameter**: It is the shortest dimension of the minimum projected area.

In defining the size the three principal dimensions namely the length, width and thickness (major, minor intermediate diameter) are measured to a given accuracy (Fig. 1.1).

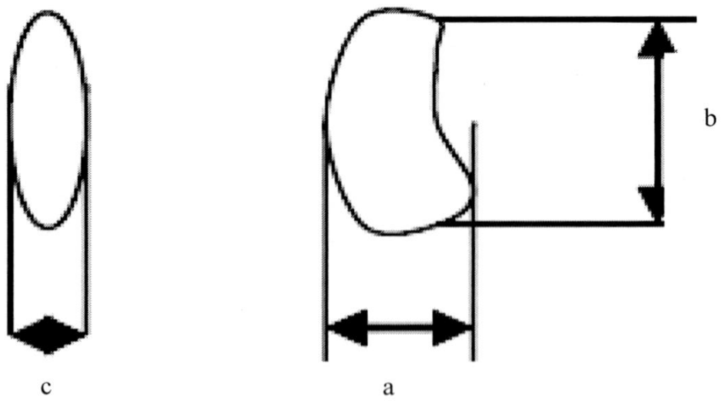

Fig. 1.1. Three principle dimensions of seed

Geometric mean diameter

The geometric mean diameter (D_g) of agricultural material is determined using following relationship:

$$D_g = (axbxc)^{1/3}$$

Arithmetic mean diameter

The arithmetic mean diameter (D_a) can be calculated using the following relationship:

$$D_a = \frac{a+b+c}{3}$$

Aspect ratio

Aspect ratio is defined as the ratio of intermediate diameter to major diameter. It is also defined as the ratio of width to length

$$\text{Aspect ratio} = \frac{\text{Intermediate diameter}}{\text{major diameter}}$$

Frontal area

The frontal area can be determined by:

$$A_f = \frac{\pi}{4}(D_g)^2$$

Slenderness ratio:

Slenderness ratio is the ratio of the effective length to the least radius of a particle and it may be calculated as:

$$\text{Slenderness ratio} = \frac{\text{minor diameter}}{\text{major diameter}}$$

Roundness

It is a measure of the sharpness of the solid material. The most important parameter to measure the size of the granular materials is roundness (Fig. 1.2). The most widely accepted method for determining the size of irregular particle is given below.

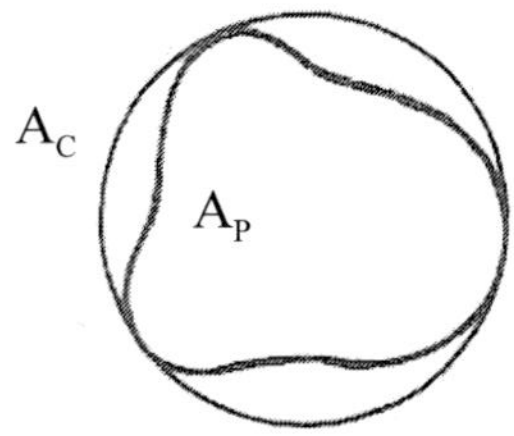

Fig. 1.2. Roundness

$$\text{Roundness} = \frac{A_P}{A_C}$$

where,

A_p = largest projected area of the particle when it is in natural rest positionA_c = area of smallest circumscribing circle

Instruments for size measurement

- The length, width, and thickness terms are commonly used that correspond to major, intermediate, and minor diameters, respectively.
- The dimensions can be measured using a calliper (Fig. 1.3) and micrometer (Fig. 1.4).

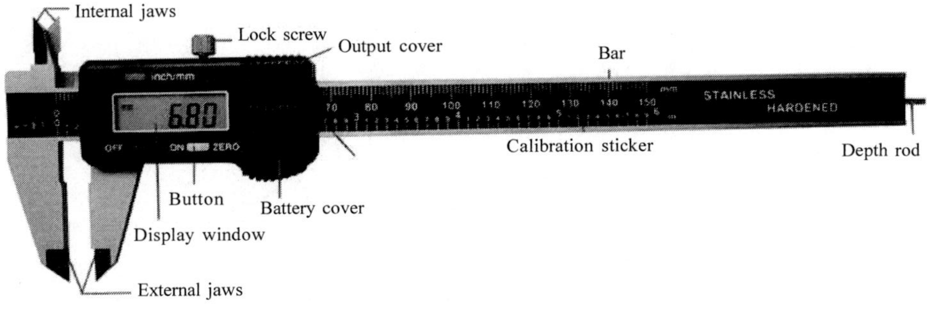

Fig. 1.3. Digital caliper for measuring dimensions of solid particulates

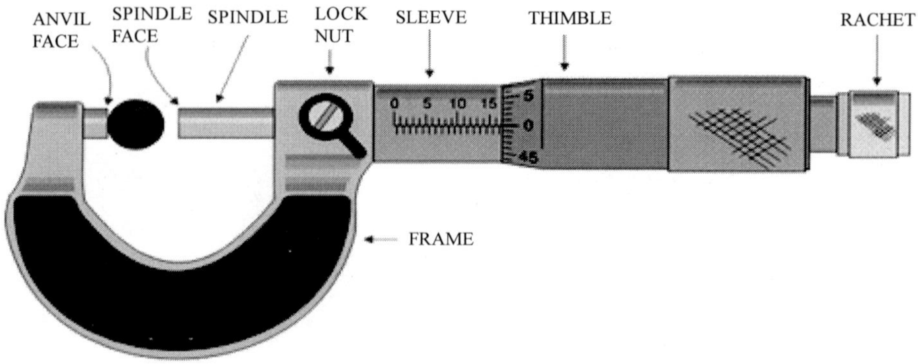

Fig. 1.4. Micrometer for measuring size of the solid materials

- The micrometer is a simple instrument used to measure distances between surfaces.
- Most micrometers have a frame, anvil, spindle, sleeve, thimble, and ratchet stop.
- They are used to measure the outside diameters, inside diameters, the distance between parallel surfaces, and the depth of holes.

Determination of shape of the food materials

Shape is important property that used in heat and mass transfer calculations, screening solids to separate foreign materials, grading of fruits and vegetables, and evaluating the quality of food materials. The shape of a food material is usually expressed in terms of its sphericity, aspect ratio, ellipsoid ratio and slenderness ratio.

Sphericity

Sphericity may be defined as the ratio of the diameter of a sphere of the same volume as that of the particle to the diameter of the smallest circumscribing sphere (Fig.1.5).

Sphericity can also be defined as the ratio of the diameter of the largest inscribing circle (D_i) to the diameter of the smallest circumscribing circle (D_c)

$$\text{Sphericity} = \frac{D_i}{D_c}$$

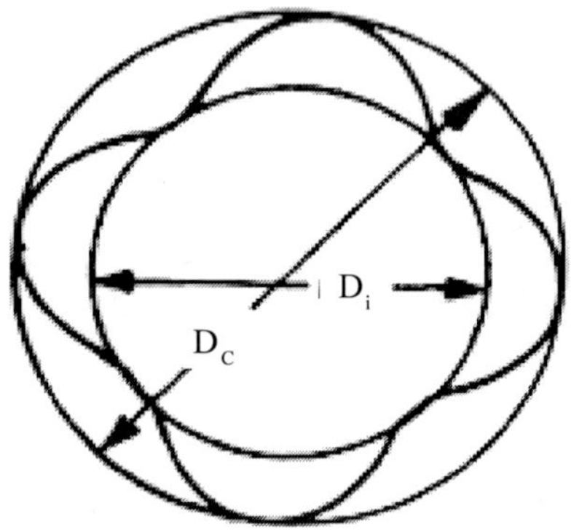

Fig. 1.5. Diagram for sphericity

Sphericity can also be given as the ratio of the surface area of a sphere of same volume (S_s) to the surface area of the particles (S_p)

$$\text{Sphericity} = \frac{S_s}{S_p}$$

Degree of sphericity

Considering the volume of the particle is equal to the volume of an ellipsoid having three axis with intercepts l, b and t (Fig.1.6). The diameter of the circumscribed sphere is the largest intercept (l) of the ellipsoid as shown in Fig. The degree of sphericity can be expressed as given below:

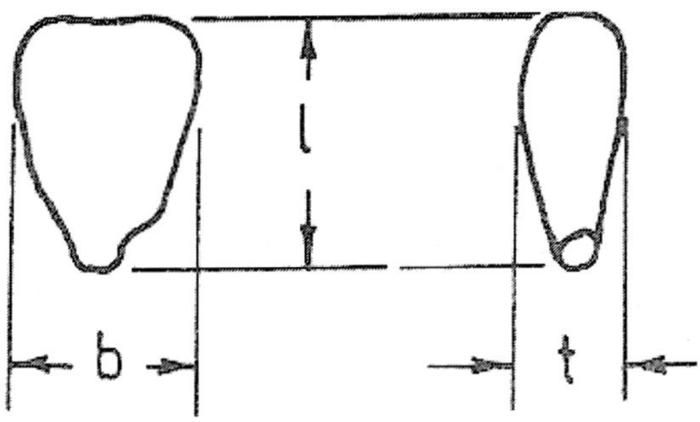

Fig. 1.6. Intercepts of elongated grain

$$\text{Sphericity} = \sqrt[3]{\frac{\text{Volume of the particle}}{\text{Volume of circumscribed sphere}}}$$

$$= \left[\frac{\frac{\pi}{6}lbt}{\frac{\pi}{6}l^3}\right]^{1/3} = \frac{(lbt)^{1/3}}{l} = \frac{\text{"geometric mean diameter"}}{\text{"major diameter"}}$$

Where,

l = largest intercept

b = largest intercept perpendicular to l

t = largest intercept perpendicular to l and b

The geometric mean diameter of the particle is also called as the ‘equivalent diameter’.

Example:

Calculate the sphericity of a cylindrical object of diameter 1.0 cm and height 3 cm.

Given: D = 1 cm; R = 0.5 cm; h = 3 cm

$$\text{Volume of particle} = \pi \times \frac{D^2}{4} \times h = 3.14 \times \frac{1^2}{4} \times 3 = 2.356\text{cm}^3$$

Radius of sphere of volume 2.356 cm^3

$$\frac{4 \times \pi r^3}{3} = 2.356$$

r = 0.8255 cm

Surface area of sphere of same volume as the particle (S_s)

$= 4\ \pi\ r^2$

$= 4 \times \pi \times (0.8255)^2$

$= 8.563\ cm^2$

Surface area of particle (S_p) $= 2\pi R\ (h + R)$

$= 2 \times \pi \times 0.5 \times (3 + 0.5)$

$= 10.966\ cm^2$

$$\text{Sphericity} = \frac{S_s}{S_p} = \frac{8.563}{10.966} = 0.779$$

Result

The size of the given food grains ____________________

The shape of the given food grains____________________

2

Determination of Bulk Density of The Agricultural Materials

Aim

- To determine the bulk density of the powder.

Materials required

- Raw material (powder)
- Measuring cylinder
- Electronic weighing balance
- Volumeter

Theory

Bulk density of the food material play an important role in many applications such as drying, storing, separation of undesirable material, determining the purity of seeds, grinding, maturity evaluations, design of storage structures, etc. It is an important parameter in packaging requirement and is often 2-20% of true density. Bulk density is the density of material when stacked or packed in bulk. Bulk density of powder can be determined under loose conditions. Loose packing is defined as the state obtained by pouring a powder sample into a vessel without any consolidation.

Bulk density

Bulk density of the food material can be defined as the ratio of weight of an untapped material to its volume including the contribution of the inter-particulate void volume. There are two kinds of voids, voids between particles and void within particles. Weight of the material can be determined by weighing balance. Volume can be determined by measured dimensions or using liquid or gas or solid displacement methods.

Unit of measurement

When we are considering the food powder, the bulk density depends on both the density of powder particles and the spatial arrangement of particles in the powder bed. The bulk density is expressed in grams per millilitre (g/mL) although the international unit is kilogram per cubic meter (1 g/mL = 1000 kg/m^3) because the measurements are made using cylinders. It may also be expressed in grams per cubic centimetre (g/cm^3).

Measurement of bulk density in a graduated cylinder

- The bulk density of a powder can be determined by measuring the volume of a known mass of powder sample that may have been passed through a sieve into a graduated cylinder.
- Pass a quantity of powder sufficient to complete the test through a sieve with apertures greater than or equal to 1.0 mm.
- If necessary, to break up agglomerates that may have formed during storage; this must be done gently to avoid changing the nature of the material.
- Approximately 100 g of the test sample (M_p) weighed with 0.1% accuracy gently introduces without compacting into a dry graduated cylinder of 250 mL (readable to 2 mL).
- Carefully level the powder without compacting, if necessary, and read the unsettled apparent volume (V_p) to the nearest graduated unit.
- If the powder density is too low or too high, such that the test sample has an untapped apparent volume of either more than 250 mL or less than 150 mL, it is not possible to use 100 g of powder sample.
- Therefore, a different amount of powder has to be selected as test sample, such that its untapped apparent volume is 150 mL to 250 mL (apparent volume greater than or equal to 60 per cent of the total volume of the cylinder); the mass of the test sample is specified in the expression of results.
- For test samples having an apparent volume between 50 mL and 100 mL a 100 mL cylinder readable to 1 mL can be used; the volume of the cylinder is specified in the expression of results.

Calculation

The bulk density can be calculated using the following formula and is expressed in g per mL or g per cm^3:

$$\rho_b = \frac{M_p}{V_p}$$

Where,

ρ_b = bulk density (g/cm^3)

M_p = mass of the powder sample (g)

V_p = volume of powder sample (cm^3)

Measurement of bulk density in a Volumeter

- Bulk density of a powder is determined by measuring the mass of a known volume of powder that has been passed through a volumeter into a cup (Fig.2.1).
- The apparatus consists of a top funnel fitted with a 1.0 mm sieve.
- The funnel is mounted over a baffle box containing four glass baffle plates over which the powder slides and bounces as it passes.
- At the bottom of the baffle box is a funnel that collects the powder and allows it to pour into a cup mounted directly below it.
- The cup may be cylindrical (25 mL volume with an inside diameter 30 mm) or cubical (16.5 mL volume with inside dimensions of 25.5 mm).

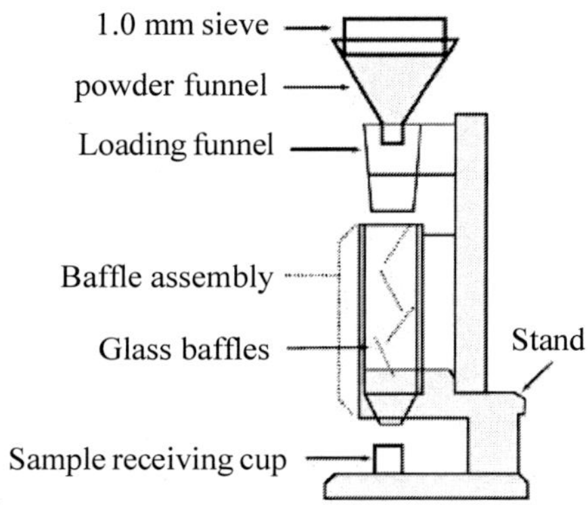

Fig. 2.1. Volumeter

Procedure

- Allow an excess of powder to flow through the apparatus into the sample receiving cup until it overflows, using a minimum of 25 cm^3 of powder with the cubical cup and 35 cm^3 of powder with the cylindrical cup.
- Carefully, scrape excess powder from the top of the cup by smoothly moving the edge of the blade of spatula perpendicular to and in contact with the top surface of the cup, taking care to keep the spatula perpendicular to prevent packing or removal of powder from the cup.

- Remove any material from the side of the cup and determine the mass (m) of the powder to the nearest 0.1 per cent.

Calculation

The bulk density can be calculated using the following formula and is expressed in g per mL:

$$\rho_b = \frac{m}{V_0}$$

Where,

ρ_b = bulk density (g/mL),

m = mass of the powder (g),

V_0 = volume of the cup (mL)

Result

Bulk density of the given powder is_______________.

3

Determination of Tapped Bulk Density and Flowability of The Food Powder

Aim

- To determine the tapped bulk density and flowability of the food powder.

Materials required

- Raw material (powder)
- Measuring cylinder
- Electronic weighing balance

Tapped bulk density of the powder

Tapped bulk density is measured for powdery samples. The volume is measured after tapping the powdery material so that the product is adjusted to inter-granular spaces, but it cannot eliminate all the inter-granular spaces. The tapped density is an increased bulk density attained after mechanically tapping a container containing the powder sample.

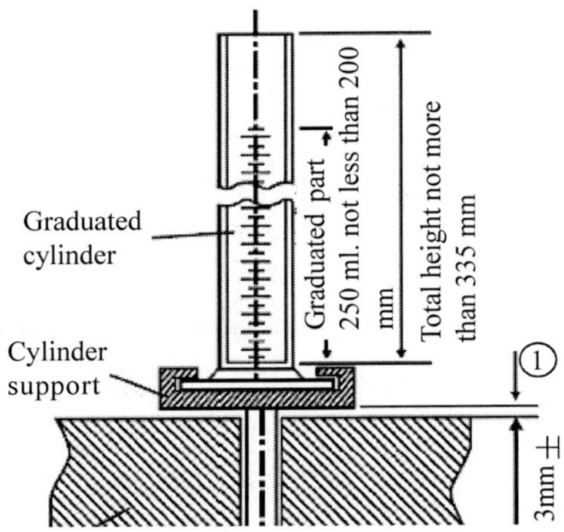

Fig. 3.1. Schematic diagram of tapped density measurement

Determination of tapped density is a method to determine the bulk density of food powder under tapped packing condition. Tapped packing is defined as the state obtained when the vessel containing the powder sample is to be repeatedly dropped a specified distance at a constant drop rate until the apparent volume of sample in the vessel becomes almost constant.

Procedure

- The tapped density is obtained by mechanically tapping a graduated measuring cylinder containing the powder sample (Fig.3.1).
- Take required quantity of the powder sample in a 50 mL volume graduated measuring cylinder and weigh the mass of the sample (g).
- After observing the initial powder volume and mass, the measuring cylinder should be tapped mechanically repeatedly from a specified height (maximum 15 cm).
- Measure the final volume of the powder sample after tapping (V_f).
- The mechanical tapping is achieved by raising the cylinder and allowing it to drop, under its own mass, a specified distance.
- Devices that rotate the cylinder during tapping may be preferred to minimize any possible separation of the mass during tapping down.
- Carry out 10, 50, 100, etc, taps on the same powder sample and read the corresponding volumes V_{10}, V_{50}, V_{100}, etc, to the nearest graduated unit. Fewer taps may be appropriate for some powders, when validated.

Calculation

Mass of the powder sample (m) = g

Final tapped volume (V_f) = cm^3

Tapped bulk density ($ñ_t$) = g/mL or g/cm^3

$$\text{Tapped density } (\rho_t) = \frac{\text{Mass of the powder, m}}{\text{Final tapped volume of powder, } V_f}$$

Measurement of flowability of the powder

- Bulking properties of a powder are influencing by the inter-particulate interactions and the interactions are interfere with powder flow.
- A comparison of the bulk and tapped densities can give a measure of the relative importance of these interactions in a given powder. Such a comparison is often used as an index of the ability of the powder to flow.

- Compressibility index and Hausner ratio are measures of the tendency of a powder to be compressed as described above.
- As such, they are measures of the powder ability to settle and they permit an assessment of the relative importance of inter-particulate interactions.
- In a free-flowing powder, such interactions are less significant, and the bulk and tapped densities will be closer in value.
- For poorer flowing materials, there are frequently greater inter-particulate interactions and a greater difference between the bulk and tapped densities will be observed.
- These differences are reflected in the Compressibility Index and Hausner ratio.

Calculation

$$\text{Compressibility index (CI)}=\frac{V_0-V_f}{V_0}\times 100$$

$$\text{Hausner ratio (HR)}=\frac{V_0}{V_f}$$

Where,

V_0 = unsettled apparent volume (mL)

V_f = final tapped volume (mL)

Alternatively

Compressibility index also called as Carr index. The bulk and tapped densities terms are used for determination of the Carr index and Hausner ratio as stated below:

$$\text{Carr index (CI)}=\frac{\rho_0-\rho_b}{\rho_t}\times 100$$

$$\text{Hausner ratio (HR)}=\frac{\rho_t}{\rho_b}$$

Where,

ρ_t = tapped density, kg/m^3

ρ_b = bulk density, kg/m^3

Table 1. Classification of powder flowability based on Carr index and Hausner ratio values.

Flowability	Carr index (%)	Hausner ratio
Excellent	0-10	1.00-1.11
Good	11-15	1.12-1.18
Fair	16-20	1.19-1.25
Passable	21-25	1.26-1.34
Poor	26-31	1.35-1.45
Very poor	32-37	1.46-1.59
Extremely poor	> 38	> 1.60

Result

i) Tapped bulk density of the given food powder is ________________.

ii) Flowability of the given food powder is ______________________.

4

Determination of Porosity of The Biological Materials

Aim

- To determine the porosity of the powder

Materials required

- Gas pycnometer
- Powder sample
- Electronic weighing balance
- Graduated beaker

Theory

Porosity is a physical property characterizing the texture and the quality of dry and intermediate moisture foods. Porosity means inter-granular space. It is the measure of void space between the materials. Porosity may be defined as the percentage of volume of inter-grain space to the total volume of grain bulk. Porosity depends on size, shape, dimension, and roughness of grain surface.

The porosity of an unconsolidated agricultural material can either be determined experimentally using the porosity tank method or theoretically from bulk and true densities of the material. Porosity can be determined from the difference between bulk volume of a porous material and its volume after destruction of all voids by compression. Porosity determination requires the knowledge of the total or apparent volume, and the void volume included in the material matrix; their difference is also known as the particle volume.

Total mass

Total mass of the material may be defined as sum of mass of dry solids and mass of water.

$$M_t = M_s + M_w$$

Where,

M_t = total mass, g

M_s = mass of dry solids, g

M_w = mass of water, g

Total volume

The material to be analyzed is made up of a solid matrix, water, and an air phase. The air phase is present in open and closed pores. The open or interconnected pores may be defined as the pores connected with each other and with the external surface of the material. Thus the total (apparent) volume of the material is defined as:

$$V_t = V_s + V_w + V_a$$

Where,

V_t = total volume of the material

V_s = volume of dry solid matrix

V_w = volume of water

V_a = volume of air in pores

Particle volume

The particle volume is the sum total volume of the sample excluding the volume of the open pores

$$V_p = V_s + V_w$$

Total porosity

Total porosity of particulate materials includes the voids within and among the particles. Total porosity may be defined as the ratio of the air volume to the total (apparent) volume.

$$\varepsilon_p = \frac{V_t - V_p}{V_t}$$

Determination of volume of a sample by gas pycnometer

A pycnometer is used to measure volume and density of solid materials in a non-destructive manner. It is working based on the Archimedes' principle of fluid displacement and Boyle's law of volume-pressure relationships. Archimedes' principle is that an object partially or totally immersed in a fluid is lifted up by a force equal to the weight of the fluid that is displaced. Ideally, a gas is used as the displacing fluid since it penetrates the finest pores, allowing maximum accuracy. That is why helium is used preferentially, since its small atomic dimension enables entry into pores approaching one Angstrom (10^{-10} m). A constant-volume gas pycnometer is composed of a sample chamber, a tank and an absolute pressure transducer, which is positioned in a thermostatically controlled environment (Fig.4.1).

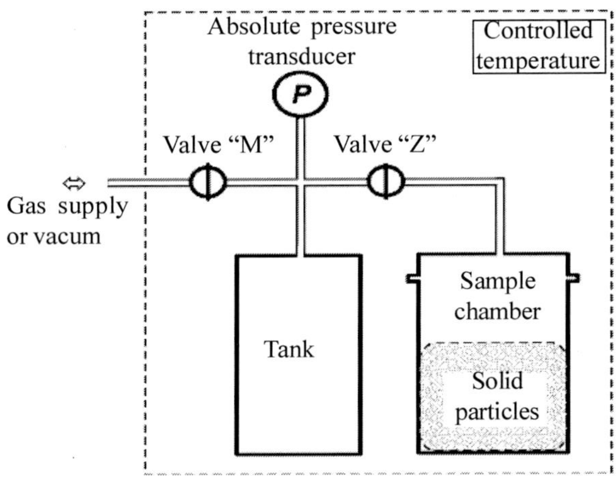

Fig.4.1. Constant-volume helium gas pycnometer

Procedure

- The sample is partially filled in the sample chamber (also known as pycnometer).
- The pycnometer is filled with gas by opening the valves 'Z' and 'M'.
- Measure the initial gas pressure in the pycnometer (P_i) by using the absolute pressure transducer after closing the valve 'M'.
- Isolate the sample chamber (pycnometer) by closing the valve 'Z'.
- Gas is introduced into the tank or removed from it by opening the valve 'M'
- The valve 'M' is closed again and measure the gas pressure in the tank (P_j).
- Open the valve 'Z' so that the gas can expand from the tank to the sample chamber (or vice versa).
- Measure the final gas pressure (P_f) in the sample chamber when the gas expansion is finished.

- Based on the hypotheses that the gas behaves ideally and that the expanding gas quickly reaches equilibrium, the following equation can be used:

$$V_s = V_c + V_t\left(\frac{P_f - P_j}{P_f - P_i}\right)$$

Where,

V_s = Volume of the sample

V_c = Volume of the sample chamber

V_t = Volume of the tank (reservoir)

P_i = Initial gas pressure in the sample chamber

P_f = Final gas pressure in the sample chamber

P_j = Gas pressure in the tank

Bulk density

$$\text{Bulk density } (\rho_b) = \frac{\text{Mass of the material,g}}{\text{Volume of the material,cm}^3}$$

True/particle density

Particle density may be defined as the ratio of total mass to true volume or particle volume.

$$\text{Particle density, } \rho_p = \frac{\text{Total mass, } m_t}{\text{Particle volume, } V_s}$$

Porosity

$$\text{Porosity} = \left(1 - \frac{\text{Bulk density}}{\text{True density}}\right) X \frac{1}{100}$$

Result

The porosity of the given powder sample is_________

5

Determination of Specific Gravity of The Agricultural Materials by Pycnometer Method

Aim

- To determine the specific gravity of the agricultural materials by Pycnometer method.

Materials required

- Pycnometer
- Vacuum pump
- Mortar and pestle
- Balance weighing 0.01 g
- Thermometer

Theory

The specific gravity is defined as the unit weight of the product mass divided by the unit weight of distilled water at 4°C. The unit density is obtained by measuring the volume and weight of a unit product. Sometimes it is required to compare the density of the product to the density of water. This comparison is in the form of ratio and is termed as the specific gravity.

Specific gravity also known as relative density is the ratio of density of a substance to the density of a standard substance at the same temperature and pressure. For solids and liquids the standard substance is water while for gasses it is air. The specific gravity of food material play an important role in design of storage bins, conveying equipments, separation equipments, etc. The specific gravity is frequently used to solve the various phase relationships, including void ratio, porosity and degree of saturation together with the moisture content and unit weight. It is also used in calculations associated with the grain size analysis, consolidation and compaction. Pycnometer method, which is a standardized vessel for determining specific gravity. It involves a standard container of accurately defined volume used to determine the specific gravity of solids and liquids.

Procedure

- Weigh empty and dry Pycnometer to the nearest 0.01 g and record the weight as W_1 g.
- Take about 100 g of oven dried sample (food powder) and transfer it into the Pycnometer.
- Weigh the Pycnometer and dry sample to the nearest 0.01 g and record the weight as W_2 g.
- Add distilled water in the Pycnometer until about it is two thirds full. Gently and carefully agitate the mixture.
- Apply a partial vacuum to the contents for 10 minutes to remove the entrapped air.
- Stop the vacuum and carefully remove the vacuum line from the pycnometer.
- Fill the pycnometer with distilled (water to the mark), clean the exterior surface of the pycnometer with a clean, dry cloth.
- Determine weight of the Pycnometer + sample + water mixture and record it as W_3 g.
- Empty the pycnometer and wash it. Then fill it with distilled water only up to the mark.
- Clean the exterior surface of the pycnometer with a clean and dry cloth.
- Determine the weight of the pycnometer and distilled water, record as W_4g.
- Repeat the above procedure three times and determine average.
- Record the temperature of sample water mixture by the thermometer.
- Specific gravity of sample is estimated by diving the weight of sample by the weight of an equal volume of water as below:

Calculations

Volume of pycnometer	= ml
Weight of empty pycometer	= W_1 g
Weight of pycnometer + sample	= W_2 g
Weight of pycnometer + sample + water	= W_3 g
Weight of pycnometer + water	= W_4 g

$$\text{Specific gravity} = \frac{a(W_2 - W_1)}{(W_4 - W_1) - (W_3 - W_2)}$$

where, 'a' is the correction factor (a =1 at 4-35°C)

Specific gravity balance method

Specific gravity of a product is also obtained by the use of a specific gravity balance. Using this balance the product is weighed in air and in water. If the solid is heavier than water, the specific gravity is computed as follows:

$$\text{Specific gravity} = \frac{W_a}{W_a - W_w}$$

If the solid is lighter than water a sinker of a known weight is attached to the product and specific gravity is determined as:

$$\text{Specific gravity} = \frac{(W_a)_{object}}{(W_a - W_w)_{both} - (W_a - W_w)_{sinker}}$$

Where,

W_a = weight of substance in air, g

W_w = weight of substance in water, g

Result

Average specific gravity of the given agricultural product is _____________.

6

Determination of Volume of The Agricultural Materials

Aim

- To determine the volume of the agricultural materials.

Materials required

- Measuring containers
- Electronic balance
- Water and sample

Theory

The volume of food material play an important role in design of storage bins, conveying equipments, separation equipments, etc. Volume of larger products such as fruits and root crops can be determined using the platform scale or water displacement method.

Procedure

- Measure the product weight in air.
- Fill the water in a container and submerge into the known weight of the product.
- Measure the weight and volume of water displaced by the sample.
- Also measure the weight of product in water.
- The volume of the product is measured as follows.

$$\text{Volume,m}^3 = \frac{\text{weight of displaced water,g}}{\text{weight density in water,g/m}^3}$$

$$\text{Volume,m}^3 = \frac{W_{p\_air} - W_{p\_water}}{\text{weight density of water}}$$

Volume of some agricultural products can be calculated using the geometrical measurements as given below:

$$\text{Volume, } m^3 = \frac{4}{3}(\pi a b^2)$$

Where,

a = length (major diameter), mm

b = width (minor diameter) of the product, mm

W_{p_air} = weight of product in air, g

W_{p_water} = weight of product in water, g

Result

The volume of some of the agricultural materials ____________.

7

Determination of Water Solubility and Absorption Index of The Food Materials

Aim

- To determine the water solubility and absorption index of given food powder.

Materials required

- Digital centrifuge machine
- Graduated measuring cylinder
- Beaker and distilled water

Theory

Water solubility index (WSI) is one of the important physical parameters of the powder that indicates the rate of dispersing or rehydration ability in a liquid medium. Water absorption index (WAI) is the weight of water absorbed per gram of dry powder sample. In particular, WSI and WAI are two indices linked to the estimation of the behaviour of the material if further processed for use as a binder, a stabilizer, or a source of protein in beverages, health and nutrition bars, dairy, baked, and emulsified/ground meat food systems. The method involves adding a known weight of powder to distilled water. The sample is then stirred and the suspension is poured into a centrifuge tube. After centrifugation, the supernatant is dried and the weight of solids in the supernatant is used to calculate the WSI as a percentage of dry weight of the powder. The weight of the sediment in the centrifuge tube is used to calculate the WAI as a percentage of the water bound per 100 g of dry powder.

Procedure

- Dissolve five gram of powder sample in 100 ml distilled water in a beaker at room temperature and stirred for 20 min.
- Centrifuge the juice concentrate at 5000 rpm for 10 min using digital centrifuge machine.

- Transfer the supernatant solution into a petri dish and dried it in a hot air oven at 130°C.
- Weigh the solids in dried supernatant and determine the water solubility index (%).
- The water absorption index is the ratio of weight of wet solids remaining in the centrifuge tube to the dry weight of initial powder sample.
- Weight of dry solids in dried supernatant measured as W_{ds} (g).
- Weight of wet solids remaining after centrifugation is indicated as W_{ws} (g).
- Weight of initial dried powder sample recorded as W_p (g).

$$\text{Water solubility index} = \frac{W_{ds}}{W_p} \times 100$$

$$\text{Water solubility index} = \frac{W_{ws}}{W_p}$$

Result

Water solubility and absorption index of the food powder are ______ and ______ respectively.

8

Determination of Angle of Repose of Biological Materials

Aim

- To determine the angle of repose of the given biological materials.

Materials required

- Angle of repose apparatus
- Grains
- Scale

Theory

Angle of repose is an important property used in particulate foods such as seeds, grains, etc. When granular solids are piled on a flat surface, the sides of the pile are at a definite reproducible angle with the horizontal. This angle is called the angle of repose of the material. The angle of repose is the angle between the base and the slope of the cone formed on a free vertical fall of the granular material to a horizontal plane. The angle of repose is important for the design of processing, storage, and conveying systems of particulate materials.

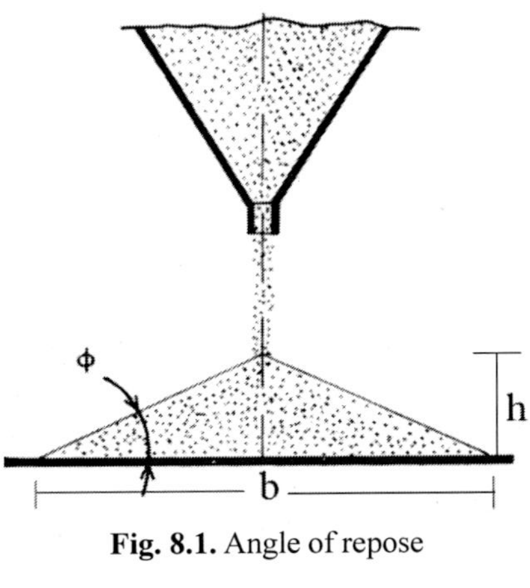

Fig. 8.1. Angle of repose

Factors

The size, shape, moisture content, surface roughness and orientation of the grains affect the angle of repose. When the grains are smooth and rounded, the angle of repose is low. For very fine and sticky materials the angle of repose is high. The angle of repose is increase with the increase of moisture content of material. This variation of angle of the repose with moisture content occurs because surface layer of moisture surrounding the particle holds the aggregate of grain together by the surface tension.

Test Procedure

- Grain sample is filled in the box or funnel that attached to the apparatus and lift up gradually at determined height.
- Allow the sample to free flow, accumulate and form a conical heap on the surface.
- Then, measure the width of base (b) and height of cone (h) using scale.
- Angle of repose is calculated from the ratio of the height to the base radius of the heap formed.

$$\theta = \tan^{-1}\left(\frac{2h}{b}\right)$$

Where,

θ= angle of repose

h = height of cone (cm)

b = width of base (cm)

Results

i) The angle of repose of the paddy is _____________.

ii) The angle of repose of the mustard is ____________.

iii) The angle of repose of the wheat is _____________.

9

Determination of Thermal Conductivity of the Agricultural Materials by Probe Method

Aim

- To determine the thermal conductivity of the agricultural materials probe method.

Materials required

- Thermal conductivity probe
- Stainless-steel sample container
- Grain or powder sample
- Graph sheet
- Power source

Theory

Thermal properties are important in design of dryers, development of thermal processing system and solving various problems on heat transfer operation. The raw foods are subjected to various thermal treatment and change of temperature depends on the thermal properties of the product. Applications of thermal treatment are stimulating germination, thermal killing of insect-pest in storage, fly quarantine disinfestations in fruits. The thermal conductivity may be defined as the rate of heat flow through unit thickness of material per unit area normal to the direction of heat flow and per unit time for unit temperature difference. Thermal conductivity is an important parameter for calculation of heat requirement in design of a dryer, calculation of sensible heat and latent heat requirement, modeling thermal processes including cooking, drying, sterilization, pasteurization, chilling, and freezing. Such models allow for process optimization to improve quality and efficiency. The enzymatic activities and attack of microorganisms of fruits and vegetables can be minimized for extending shelf life by lowering the storage temperature.

Description of thermal conductivity probe

Thermal conductivity probe is used for measurement of thermal conductivity of most of the food materials because of its relative simplicity and speed of measurement. Thermal conductivity measurement is not easy because it depends on the structure and chemical composition of the food materials. Thermal conductivity probes have three practical sizes viz. miniature, small, and large or field probes. Miniature probes are often built to fit within a hypodermic needle and these probes are particularly suited to measurements on small samples, biological materials, foodstuffs, etc. A typical line heat-source thermal conductivity probe is shown in Fig.9.1. A typical probe consists of a 20-gauge hypodermic needle of 0.9 mm diameter and 45 mm long connected to a handle. An insulated chromel constantan thermocouple wire of 0.051 mm diameter is emplaced in the hypodermic needle and wrapped around the upper half of the probe handle. A constantan heater wire of 0.077 mm diameter insulated with plastic spaghetti tubing is also emplaced in the hypodermic needle. The purpose of using constantan wire is to avoid the change in electrical resistance with temperature. The wire is taken from the handle to the top and back.

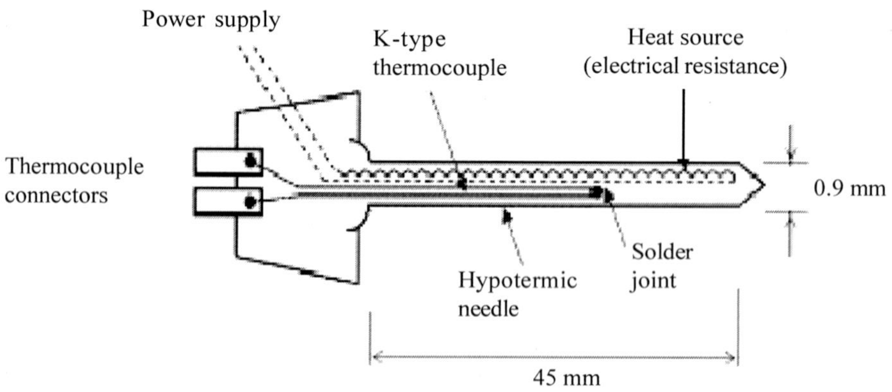

Fig.9.1. Typical line heat-source thermal conductivity probe

Procedure

- Take sample (grains or powders) in a watertight, stainless-steel sample container and shake and tap several times until the sample is settle down.
- The sample container in placed in a constant temperature controlled bath and equilibrated at room temperature.
- Perform the measurement with base-line temperature (record initial temperature).

- The thermal conductivity probe is inserted into the sample having uniform initial temperature, which is allowed to equilibrate at the base-line temperature.
- The heater is switched on and allow the sample to reach equilibrium.
- The probe is heated at a constant rate of energy input, and record the probe temperature.
- The heater is switched off after a period of time (2-4 min)
- Then allow the sample to return to the base-line temperature before a replicate measurement.
- Plot a slope of natural logarithm of time versus temperature adjacent to the line heat source, the slope is equal to $Q/4\pi K$.
- Determine the value of the slope of the linear portion of the curve.

Calculation

Thermal conductivity of the sample can be estimated by the following formula:

$$K = Q \times \frac{\ln\left[(t_2 - t_0)/(t_1 - t_0)\right]}{4\pi(\Delta T_2 - \Delta T_1)}$$

Where,

K = thermal conductivity of the sample, W/m°C

Q = power generated by the probe heater, W/m

t_1 and t_2 = time during the linear portion of the ΔT versus ln*t* curve, s

ΔT_1 and ΔT_2 = correspond to the differences between the probe thermocouple and base-line temperatures at time t_1 and t_2 respectively, °C

Result

The thermal conductivity of the given sample is__________

10

Determination of Specific Heat of The Agricultural Materials by Differential Scanning Calorimetry Method

Aim

- To determination the specific heat of the agricultural materials by differential scanning calorimetry method.

Materials required

- Differential scanning calorimetry
- Power supply
- Computer
- Sample

Theory

Specific heat is defined as the amount of heat needed to raise the temperature of one kg of the material by one degree at constant pressure without a change in state. With food materials, specific heat is a function of different components of a food, moisture content, temperature, and pressure. Specific heat is an important parameter in thermal analysis of food processing and equipment used in heating or cooling of foods. The numerical values of specific heat of the food materials are used in designing food processes and processing equipment. It can be used to calculate the heat load imposed on the refrigeration equipment by the freezing of foods. An apparent specific heat is used in freezing or thawing and in most food processing applications.

The specific heat of a food increases as the product moisture content increases. In unfrozen foods, specific heat capacity is relatively constant with respect to temperature. However, for frozen foods, there is a large decrease in specific heat capacity as the temperature decreases. The specific heat of a material can be determined experimentally by static (adiabatic) calorimetry or differential scanning calorimetry. It can be also predicted quite accurately with the help of a number of empirical equations. The equations are empirical expressions, obtained by fitting experimental data into mathematical models. The mathematical models are based on one or more constituents of the food. Since water is a major component of many foods, a number of models are expressed as a function of water content.

Differential scanning calorimetry

Differential scanning calorimetry (DSC) is a thermo-analytical technique in which the temperature difference between the sample and the reference material, and amount of heat required to raise the temperature of the sample and reference materials are measured as a function of temperature. The temperature of the sample and reference materials are maintained at nearly equal throughout the experiment. Generally, the temperature program for a DSC analysis is designed such that the sample holder temperature increases linearly as a function of time. The reference sample should have a well-defined heat capacity over the range of temperatures to be scanned.

Fig.10.1 shows the schematic diagram of heat flux type DSC. It comprises the sample and reference pan, heat resistor, heat sink, and heater. The heater is supplied heat into the sample and reference material through the heat sink and heat resistor. Heat flow is proportional to the heat difference of heat sink and holders. Heat sink has the enough heat capacity as compared to the sample. In case, the sample occurs endothermic or exothermic phenomena such as transition and reaction, this endothermic or exothermic phenomena is compensated by heat sink. An exothermic process releases heat, causing the temperature of the immediate surroundings to rise. An endothermic process absorbs heat and cools the surroundings. Thus the temperature difference between the sample and the reference is kept constant. The difference the amount of heat supplied to the sample and the reference is proportional to the temperature difference of both holders. By calibrating the standard material, the unknown sample quantitative measurement is achievable.

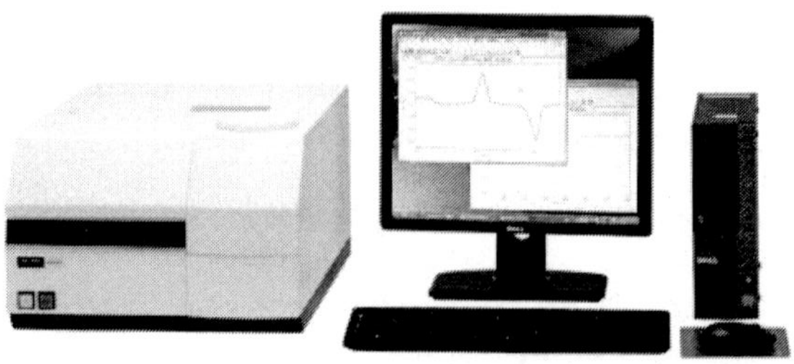

Fig. 10.1. Differential scanning calorimetry (www.hitachi-hightech.com)

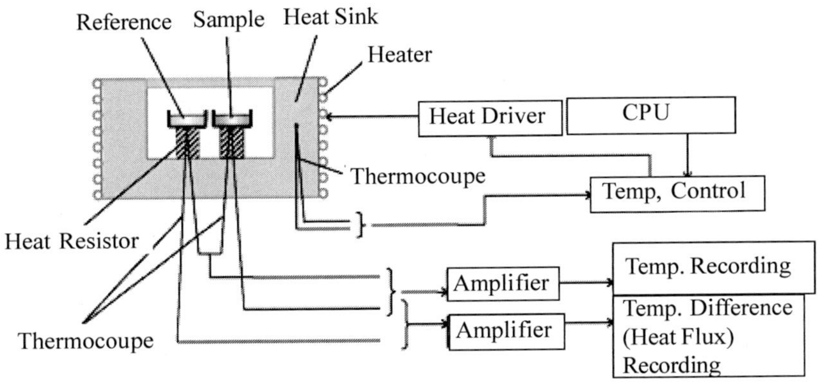

Fig. 10.2. Schematic diagram of heat flux type differential scanning calorimetry

Procedure

- Take sample and reference material in their respective pans.
- Place the sample and reference pans in the differential scanning calorimetry
- Start the experiment by switch on the DSC
- A computer is monitor the temperature and regulate the heat flow.
- The heat is supplied to the heaters at specific rate usually 10°C per minute.
- The heating rate is exactly same throughout the experiment, however the sample pan will take more heat to keep the temperature of the sample pan increasing at the same rate as the reference pan.
- Thus the heater under the sample pan need to work harder than the heater under the reference pan.
- The temperature raise can be measured by measuring the heat output of the heaters.
- Plot a graph, temperature raise on the X-axis and difference in heat output on the Y-axis.

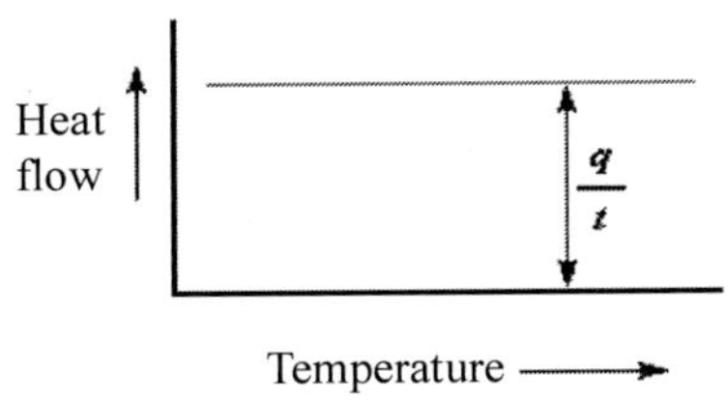

Heat flow is the amount of heat supplied per unit time to increase temperature

$$\text{Heat flow} = \frac{\text{heat}}{\text{time}} = \frac{q}{t}$$

$$\text{Heating rate} = \frac{\text{Temperature raise}}{\text{time}} = \frac{\Delta T}{t}$$

Heat capacity can be determined by dividing the heat flow by the heating rate

$$\text{Heat capacity, } C_p = \frac{\text{heat flow}}{\text{heating rate}} = \frac{q}{\Delta T}$$

Therefore, when we supply a certain amount of heat into material, its temperature will raise by a certain amount, and the amount of heat it takes to get a certain temperature increase is called the heat capacity. The heat capacity is the ratio of heat supplied to the resulting temperature increase.

Result

The specific heat of given agricultural material is ____________.

11

Estimation of Calorific Value of Agricultural Materials by Bomb Calorimeter

Aim

- To estimate the calorific value of agricultural materials using bomb calorimeter

Materials required

- Bomb calorimeter
- Food sample and water
- Electronic balance

Theory

The energy available in food is expressed in unit is called joules or calories. One kilocalorie is equal to 4.184 kilojoules. The kilocalorie is defined as the amount of heat required to raise the temperature of 1 kg of water by 1°C. The calorific/ energy value of foods is usually determined using the instrument called bomb calorimeter. It is working based on the principles of calorimetry. Basically, this device burns a food sample and transfers the heat into a known mass of water. It measures the heat produced when the food sample is ignited by an electric spark in the presence of oxygen and platinum which acts as a catalyst. The quantity of food sample to be tested is placed inner part and known volume of water is taken in outer part of the bomb calorimeter. When the food sample is electrically ignited, the surrounding water absorbs the heat produced. The energy value of the food is calculated by measuring the rise in temperature of water.

Construction of bomb calorimeter

The bomb calorimeter (Fig.11.1) consists of a heavy steel bomb, with a platinum dish and a cover held tightly in place by means of a strong screw collar, calorimeter vessel, an outer double walled water jacket, control unit to switch on/off the stirrer and ignition device, a Beckman type thermometer, a magnifying glass, and charging unit with pressure gauges to facilitate the charging of the calorimeter with oxygen. The calorimeter vessel and outer jacket wall are made in stainless steel. The bomb chamber is a container made of stainless steel that can support high pressures. The bomb is charged with pure oxygen gas through the filling valve. This bomb is introduced inside a calorimeter vessel made of stainless steel that is filled with water, and at the same time it is introduced inside a double walled water jacket. The rod of calorimeter supports a metallic crucible.

The bomb calorimeter, which contains the food sample to be burned, is hermetic to the gas by closing the filling valve and its cover. The top of the double walled jacket is closed with a cover. A Beckman thermometer is used to measure the temperature of the calorimeter vessel, which is passed through one of these orifices. The orifices also used to insert the electric wire that supplies electric current to the rod. The control unit switches on/off the stirrer and ignition device through the heating up of the thin electric wire. A load unit with pressure gauges is used to make filling with oxygen of the calorimeter easier. A magnifying glass is used to magnify the Beckman's thermometer reading accuracy.

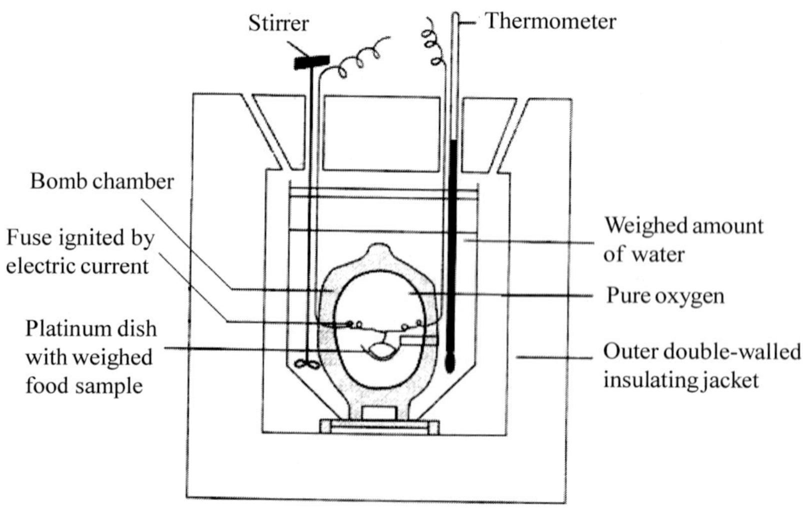

Fig.11.1. Schematic diagram of the bomb calorimeter

Working of bomb calorimeter

- A weighed amount of food sample, usually pressed into pellet form, is placed in a Platinum dish within the bomb which is then closed except for the oxygen valve, charged with oxygen to a pressure of about 300 pounds to a square inch and the oxygen valve is then closed.

- The bomb is then immersed in a weighed amount of water in the calorimeter vessel and outer double-walled jacket.
- The water in the calorimeter vessel is constantly stirred with a stirrer driven by a small motor.
- The temperature of water is measured at intervals of one minute by means of a differential thermometer, capable of being read to one thousandth of a degree.
- After the temperature of the water has been measured, the sample is ignited by means of an electric fuse.
- On account of the large amount of oxygen present, it undergoes rapid and complete combustion and generates heat.
- The heat liberated is absorbed by the water in which the bomb is immersed resulting in rise in temperature which is accurately determined.
- The energy value of the food is calculated by measuring the rise in temperature of water.
- From the weight of the food sample and temperature rise of the water, the calorific value can be calculated.
- The calorific value obtained in a bomb calorimeter test represents the gross heat of combustion per unit mass of fuel sample.
- This is the heat produced when the sample burns, plus the heat given up when the newly formed water vapour condenses and cools to the temperature of the bomb.

Result

The calorific value of given food materials is__________

12

Determination of Texture Profile of The Different Food Materials

Aim

- To determine the texture profile of the food materials.

Materials required

- Texture analyzer
- Food materials

Theory

Texture testing is a well-established technique for evaluating the mechanical properties of raw materials, food structure, pre- and post- harvest quality control checks. The texture testing has many applications across wide range of foods including cereals, fruits, vegetables, meat, poultry, fish, baked foods, confectionaries, snacks, dairy, gelatins, pasta, etc. Food texture analysis of foods can highlight the quality improvement throughout the supply chain. At research and development stage, new or alternative ingredients can be compared with existing ingredients.

Texture profile parameters

The texture profile parameters including hardness, firmness, crispiness, rupture strength springiness, chewiness, cohesiveness, gumminess, gel strength, adhesiveness, consistency, crunchiness, elasticity, extensibility, fracturability, stiffness, stringiness, toughness, work to cut, penetrate and shear can be measured by the mechanical food texture analyzer.

i) **Hardness or Firmness:** The maximum force required to deform or penetrate or cut out the food material, expressed in Newton. The compressive force depends on the hardness or softness of the food tissues and is directly proportional to the hardness.

ii) **Cohesiveness:** Measurement of the rate at which the material disintegrates under mechanical action. It is defined as the ratio of positive force during second compression to positive force during first compression, dimensionless.

iii) **Springiness:** The ratio of distance recovered by the sample between the first and second compressions, expressed in mm.

iv) **Chewiness:** Energy required to chew a solid food to the point required to swallow it. Chewiness is calculated by multiplication of hardness × cohesiveness × springiness, expressed in Newton.

v) **Resilience:** It measures how the sample recovers from compression relative to the distance compressed and the speed with which it is compressed.

vi) **Gumminess:** This is a product of two properties and is applicable to semi-solid products.

Types of texture analyzers

Different types of testing instrumentation from manual and motorized firmness testers to fully software-controlled texture analyzer are available.

1. Grain hardness tester

The grain hardness tester (Fig.12.1) is specially used for testing of the hardness of grains (cereals, pulses, oilseeds, etc.). Each seed is placed between the compression plates. The seed is compressed at a constant deformation rate. The applied forces at bio-yield and corresponding deformation for each seed sample is read directly from the force-deformation curve. The mechanical behaviour of seed is expressed in terms of force required for maximum strength of the seed, energy required to deform the seed to initial rupture and seed specific deformation. The rupture force is determined as the force on the digital display when the seed under compression makes a clicking sound. Each process is often completed whenever the break point of the positioned seed is reached.

Fig. 12.1. Analog grain hardness tester (Pharma test company)

2. Fruit firmness analyzer

A manual mechanical firmness tester (Fig.12.2) is suitable for determining the firmness of fruit and vegetables. It uses a portable digital force gauge to make compressive measurements. The digital force gauge is used to determine the peak forces encountered in order to test the firmness characteristics of the sample. The tester can be driven with a hand wheel and a digital indicator is used to measure the puncture depth.

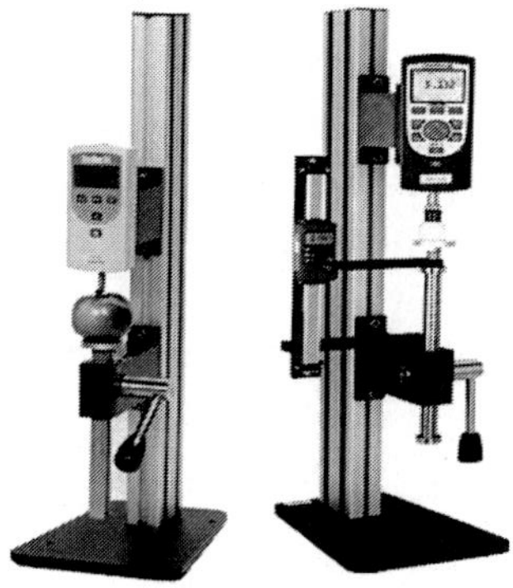

Fig.12.2. Fruit firmness analyzer (AMETEK)

3. Automated texture analyzer

The texture analyzer (Fig.12.3) is a high performance food texture analyzer for fully automated analysis includes base instrument, drip tray, food testing base, and half-inch probe for routine applications. The closed-loop drive system enables precise instrument control during a test. This permits accurate sample height measurement for testing samples to a percentage of their original size, with the ability to adjust test speed relative to sample dimensions. This analyzer is configured by software fundamental algorithms.

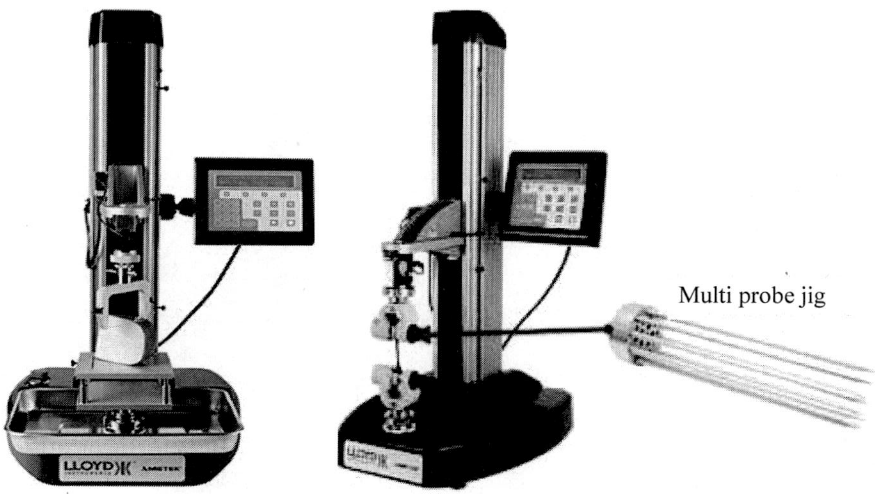

Fig. 12.3. Automated texture analyzer (AMETEK)

A wide range of probes fixtures and jigs including cones, cylinders and ball can allow use of a single instrument to make a diverse range of measurements in a variety of foods. The choice of probes/fixtures depends on the sample. If the sample has a flat surface, then generally compression platens that are larger than the sample are used. For uneven surfaces such as fruit and vegetables, small-diameter probes are used. The sample is positioned on the lower tooth and measured peak force required to bite through the sample. The choice of specimen size depends on the homogeneity of the sample. A food with large voids will require a larger sample size. In addition to the powerful measurement facilities, the software provides capabilities that can greatly enhance the presentation and understanding of test results. The video images of the entire test can be recorded and synchronized with the stress/strain data and replayed for detailed post-test analysis and still images can be taken at specific points during the test. These still images are recorded on the graph for ease of analysis.

4. Compression tester

Food compression tests are one of the most common tests carried out on food products, reflecting the action of the human jaw. Typically a sample is placed on the flat base of the instrument and an upper compression platen is lowered onto the sample. The platen should be bigger than the sample being tested, otherwise it becomes a puncture test. Tests are performed by compression to a given force, a given position or a percentage of the original height of the sample. Texture profile analysis is widely used based on double compression of a material between two flat plates. For bread, it is widely used because it mimics the biting and chewing action of the jaw by compressing the sample, allowing it to recover briefly and then compressing again (Fig.12.4).

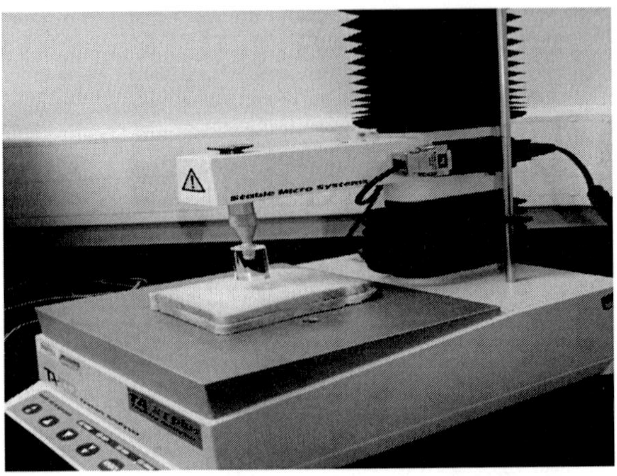

Fig. 12.4. Compression tester (Campden BRI)

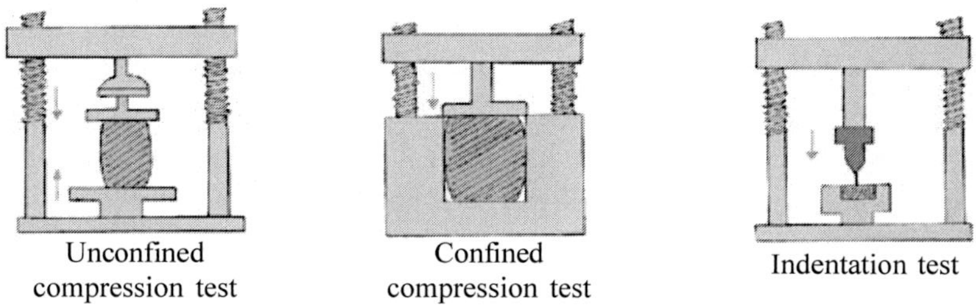

The force-time plot of the test appears as two peaks, representing the two compressions. The second peak is typically smaller than the first, since the crumb structure will be weakened by the first compression and will be less resistant to the second compression. The gap between the two peaks represents the time when the probe is stationary between the two compressions. The test is usually performed on a cylindrical core taken from the centre of a slice of bread. The height of the first peak is the key feature of the trace and is related to the firmness or hardness of the bread. This parameter shows good correlation with the sensory evaluation of freshness for bread. Other attributes measured include resilience, cohesiveness, and springiness, as shown in Fig.12.5. This method is also suitable for measuring changes in firmness during storage.

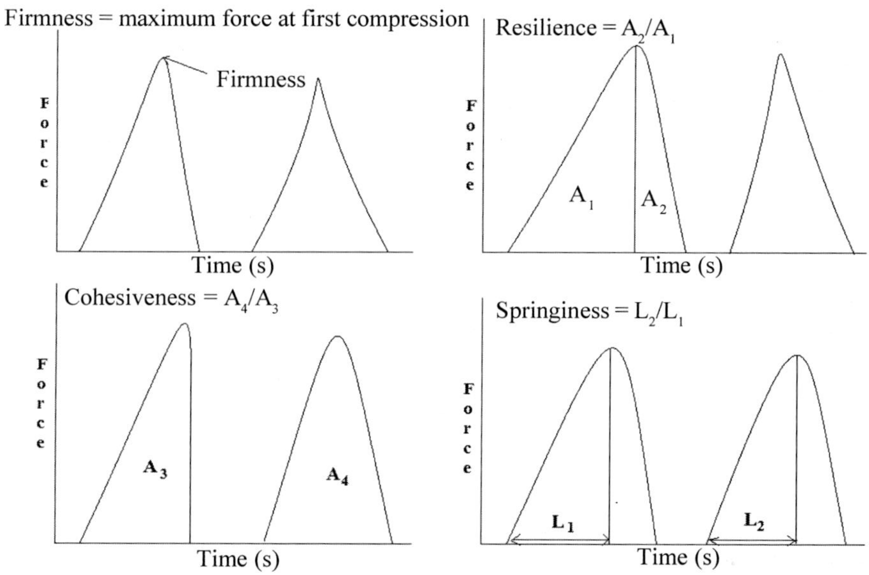

Fig. 12.5. Measures of firmness, resilience, cohesiveness and springiness (Campden BRI)

5. Shear tester

Shear testing is appropriate because many foods are first cut or sheared by the front incisors when introduced to the mouth. There are many different variations on the basic shear testing apparatus such as razor blades, V shaped blades, rounded blades and straight blades. A specific version called the Warner-Bratzler shear force test is widely used in the meat industry to measure tenderness of meat (Fig.12.6). It consists of a stainless steel blade in which a hole consisting of an equilateral triangle is cut; the point of the triangle rounded off to specific radius.

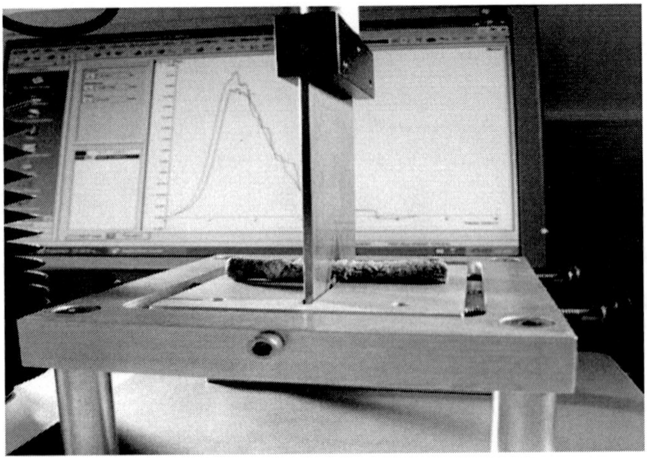

Fig. 12.6. Warner-Bratzler method for meat texture (Campden BRI)

A cylindrical piece of meat is placed between the notch of the blade, which then moves down, forcing the meat into the "V" of the triangle until it is cut right through. The maximum force generated during the test is related to the tenderness of the meat, with a high value suggesting a tougher eating texture.

13

Determination of Moisture Content of Food Materials By Hot-air Oven Method

Aim

- To determine moisture content of given agricultural materials by hot-air oven method

Required materials

- Hot-air oven
- Moisture dishes or aluminium containers
- Electronic weighing balance
- Scoop, other small tools as appropriate
- Food grains

Theory

The amount of moisture in a product is given on the basis of the weight of water present in the product and is usually expressed in percent. Moisture content is designated by (i) wet basis (wb) and (ii) Dry basis (db). The moisture content on dry basis is more simple to use in calculation as the quantity of moisture present at any time is directly proportional to the moisture content on dry basis.

i) **Wet basis**: The moisture content in wet basis can be represented as follows:

$$MC\%(wb) = \frac{W_m}{W_p} x 100$$

ii) **Dry basis**: The moisture content in dry basis can be determined by following expression:

$$MC\%(db) = \frac{W_m}{W_d} x 100$$

Where,

W_m = Weight of moisture in products (g)

W_p = Weight of product sample (g)

W_d = Weight of dry matter of product sample (g)

Relationship between dry basis and wet basis moisture content

The value of dry basis moisture content is more than the wet basis moisture content. The relationship between dry basis and wet basis moisture content is given by the following expression:

$$MC(db)\% = \frac{MC(wb)}{100 - MC(wb)} x 100$$

For determination of moisture content of a particular product, the choice of method depends on many factors, they are (a) the form in which water is present in the product, (b) the relative amount of water present, (c) the rapidity of determination, (d) accuracy of method, (e) product's nature whether easily oxidized or decomposed and (f) the cost of equipment used. Direct methods are very simple and accuracy of moisture content is high, hence these methods are used by researchers. The direct methods are time consuming therefore such methods are not very much useful for determination of moisture in field, mandi or ware houses.

Test procedure

The air-oven drying method can be accomplished in a single stage or double stage in accordance with the grain samples containing less than 13% or more than 13% moisture content.

a) **Single stage method**

When the moisture content of grains is less than 13%, 2 - 3 grams representative ground samples of grains are placed in an air-oven. The temperature of the oven is set at 130°C and the samples are kept in oven for 1-2 hours. Afterwards, the samples are taken out and placed in a desiccator to cool down. The drop in the weight of grain is measured based on its initial weight.

b) **Double stage method**

If the moisture content of grains more than 13%, 25 to 30 grams of whole unground representative samples of grains are taken and placed in an air-oven at 130°C for 14-16 hours so that its moisture content is reduced to about 13 per cent. Afterwards, the samples are taken out from oven placed in a desiccator to cool down to room temperature. Then follow the same procedure as in single stage method.

Observation and calculation

W_1 = weight of sample before drying, g

W_2 = weight of dried sample after drying, g

W_m = Weight of moisture in the sample ($W_1 - W_2$), g

$$MC\%(wb) = \frac{M_m}{M_1} \times 100$$

$$MC\%(db) = \frac{M_m}{M_2} \times 100$$

Result

Moisture content of given food material by hot-air oven method is_______

14

Determination of Moisture Content of Food Grains by Digital Moisture Meter

Aim

- To determine moisture content of given food grains using digital moisture meter

Required materials

- Digital moisture meter
- Scoop
- Food grains

Theory

Indirect methods are based on the measurement of electrical properties of the grain that depends upon moisture content. There are two indirect methods namely (i) electrical resistance and (ii) dielectric methods are used. The electrical conductivity or resistance of a product measures its moisture content at a given compaction and temperature. The electrical resistance is depends on its moisture content, degree of compaction, grain temperature and impurities present in the sample. The universal digital moisture meter measures the electrical resistance of the grain at a given compaction. Universal moisture meter gives fairly accurate readings of moisture content on wet basis. In dielectric method, grain moisture content can be measured with the help of dielectric properties of the grains.

Digital moisture testing machine

The digital moisture testing machine has been designed for quick and accurate determination of percent moisture content of wide range of granular substances such as food grains, pulses, oil seeds, vegetable seeds, tea, jute, etc.

Principle

Grains are conductive to electricity due to moisture contained in them. Electrical conductivity also varies with the temperature. These properties of the grains have been put into practical use in this moisture testing machine. A temperature sensor along with Microprocessor circuitry automatically compensate for these variations and displays moisture percentage. For uniformity, accuracy and reliability of the results, the material under test is compressed in order to nullify the porosity effect of the grains.

Construction

The digital moisture testing machine (Fig.14.1) is a self contained and portable instrument. The mechanical part consists of a compression unit to compress the sample to a predetermined thickness. The thickness setting is very easily read on a vertical/circular scales. The material under test is taken in a test cup and is compressed to predetermined thickness. Press the push type switch till the reading comes in the display. No temperature reading and no correlate dial are required. Computerized version of digital moisture testing machine automatically compensates for the temperature corrections and automatically makes necessary computation for the crop. The code number can be selected from the thickness chart. It can be set on the digital moisture testing machine using thumb wheel switch.

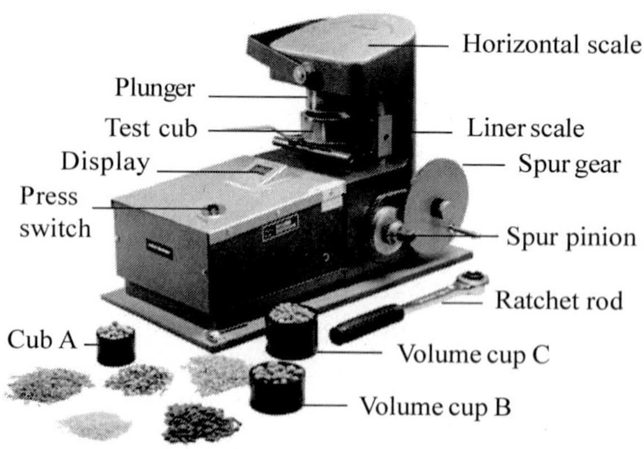

Fig.14.1. Universal digital moisture meter

Preliminary adjustment

1. Check zero setting of the scales before using the moisture testing machine.
2. Place the test cup in its housing and adjust the plunger till it touches the inner bottom of the test cup.
3. Vertical scale may be adjusted by pressing the pointer upwards or downwards as necessary.
4. Circular scale may also be adjusted by unscrew the top hexagonal screw and lift the scale, adjust as necessary and tighten the hexagonal screw.

Test procedure

1. Refer to thickness setting chart and identify the commodity to be tested. Select the commodity code for that particular commodity. Set the commodity code to required commodity code.
2. Take a representative sample in volume cup A, B, or C or by weight as specified in the thickness setting chart as per the manufacture recommendation.
3. The sample is then transfer into the test cup.
4. Place the test cub containing the sample in the lower housing of the compression unit.
5. The cup is so placed that its bottom key fits in the base socket and the lower surface of the cup settles in the cup-mounting place.
6. Lower the plunger by rotating the hand wheel till the surface of the sample touches the plunger.
7. Compress the sample to the specified thickness with the help of a Ratchet handle provided for this purpose.
8. The thickness can be read on the Linear Vertical Scale and Circular Scale.
9. Press the Switch and count-down will start from 10 and the reading will be displayed after 1 (final beep).
10. Press the switch till final beep sound. Note down reading through display unit.
11. Release the pressure and take out the test cup from its housing.
12. Remove the tested grains of the sample and pur back the test cup in housing after cleaning with the help of a soft brush or a piece of cotton cloth.
13. Now for applying high pressure the ratchet spanner is used and the sample is compressed to proper thickness.

Error conditions

1. Under range or below 3.5%, the sample is not properly pressed.
2. Repeat 3-4 times, if error persists. Try different sample and use mains instead of battery.
3. Over range, above 40%, plunger is shorted with test cup.
4. Lowbat red indicator is glowing, change the batteries or check main input.

Precautions

1. Before using the instrument check the zero settings.
2. Reuse of compressed sample is not recommended.
3. Applying pressure higher than specified, may damage the instrument.
4. Please do not touch the test cup or its housing to avoid unpleasant shock while pressing the press switch.
5. The % moisture reading depends on the temperature of the instrument.

Result

The moisture content of the given food grains is _______________.

15

Determination of Moisture Content of Biological Materials by Distillation Method

Aim

- To determine the moisture content of biological material by volumetric method.

Materials required

- Brown-Duvel fractional distillation apparatus
- Mineral oil
- Grains

Theory

This type of moisture measurement is called brown-duvel fractional distillation method. The method is recognized as an official method for determination of moisture content. In this method, directly measures the volume of moisture in cubic centimeter. This apparatus consists of flask, burner, thermometer, condenser, graduated cylinder. This apparatus consists of flask in which grain sample is boiled with mineral oil.

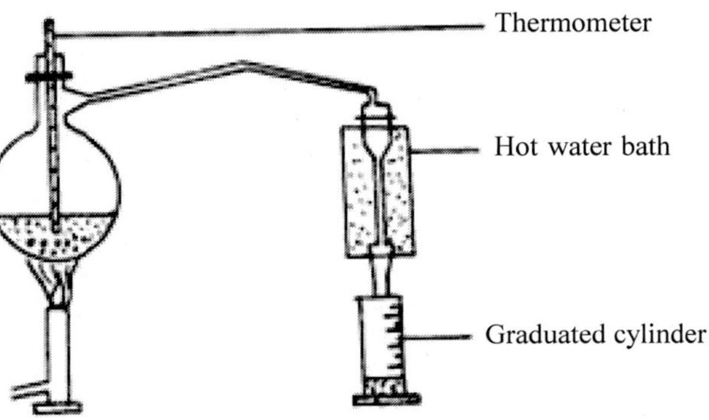

Fig. 15.1. Brown-Duvel fractional distillation apparatus

The moisture from the sample is evaporated in the form of water vapour and passed the condenser. A thermometer attached in the flask is used to measure the temperature of boiling water. The water vapour is condensed in condenser, collected in a graduated cylinder and water content is measured in milliliter.

Procedure

- Hundred grams of whole grains along with 150 ml of mineral oil is taken in a flask.
- The sample is boiled at temperature of 200°C for 30 to 40 minutes.
- Moisture from the sample is thus evaporated, condensed at condenser is collected in a graduated measuring cylinder.
- The milliliter of moisture collected shows the percentage moisture content. The moisture content is measured on wet basis.

Result

Moisture content of the given material is _______________.

16

Numerical Problems on Grain Moisture Content and Equilibrium Moisture Content

Aim

- To study the numerical problems on grain moisture content and equilibrium moisture content.

Problem 1

Five hundred kilograms of paddy at 22% moisture content (wb) is dried to 14% moisture content (wb) for milling. Calculate the amount of moisture removed in drying.

Method I

Using wet basis moisture content data:

Amount of initial moisture content $= 500 \times 0.22 = 110$ kg

Dry matter in paddy $= 500 - 110 = 390$ kg

Amount final moisture content in 14 %

$$\text{Moisture content, \%} = \frac{\text{Weight of moisture in product } (W_w)}{\text{Weight of water } (W_w) + \text{weight of dry matter } (W_d)}$$

$$0.14 = \frac{W_w}{W_w + 390}$$

Rearrange and simplify, we get,

$$Ww = \frac{390 \times 0.14}{0.86} = 63.488\text{kg}$$

Thus amount of water removed $= 110 - 63.49 = 46.51$ kg

Method II

Amount of water removed by analysis on dry basis change the initial and final moisture content to dry basis.

Amount of initial moisture content = $500 \times 0.22 = 110$ kg

Dry matter in paddy = $500 - 110 = 390$ kg

$$\text{Initial moisture content (db)} = \frac{0.22}{1\text{-}0.22} \times 100 = 28.2\%$$

$$\text{Final moisture content (db)} = \frac{0.14}{1\text{-}0.14} \times 100 = 16.27\%$$

Amount of water removed = Initial dry matter × Difference in dry basis moisture content

= $390 \times (0.282 - 0.1627) = 46.51$ kg

Problem 2

Two tones of paddy with 23 per cent moisture content on wet basis are to be dried to 12 per cent moisture content on dry basis. Calculate the weight of bone-dry products, water evaporated and dried products.

Solution

Weight of bone-dried sample = $2000 - (2000 \times 0.23) = 1540$ kg

Moisture content on dry basis for 23 % moisture content on wet basis:

$$\text{MC,\% db} = \frac{23}{100\text{-}23} \times 100 = 29.87\%$$

Therefore, water evaporated = $1540 \times (0.2987 - 0.12) = 275.2$ kg

Amount of dried product = $2000 - 275.2 = 1724.8$ kg

Problem 3

Determine the quantity of parboiled paddy with 40 % moisture content on wet basis required to produce 3000 kg of paddy with 12 % moisture content on wet basis. Work out the problem on wet basis and check the answer using dry basis.

On wet basis

Weight of paddy with 12 % moisture on wet basis = 3000 kg

Weight of bone dry paddy = $3000 - (3000 \times 0.12)$

= 2640 kg

Let,

X = amount of water present in the paddy with 40 % moisture content

Therefore,

$$\frac{40}{100}=\frac{X}{2640+X}$$

$$100X=40X+(2640\times40)$$

$$X=\frac{2640\times40}{60}=1760\text{kg}$$

Therefore, quantity of paddy with 40 % moisture content on wet basis = 2640 + 1760 = 4400 kg.

On dry basis

$$40\ \%\ \text{moisture content}=\frac{40}{100\text{-}40}=66.67\%$$

$$12\ \%\ \text{moisture content}=\frac{12}{100\text{-}12}=13.65\%$$

Amount of moisture evaporated = 2640 × (0.6666 – 0.1365) = 1399.464 kg

Total initial weight of paddy = 3000 + 1399.464 = 4400 kg

Problem 4

Determine the equilibrium moisture content of Sorghum at relative humidity of 10%, temperature of 60°C using Henderson's equation where C = 6.12×10^{-5} and n = 2.31

Henderson's equation:

$$1\text{-RH}=\exp[-CTM_n^e]$$

Where,

RH = Relative humidity, decimal

M_e = Equilibrium moisture content, % (db)

T = Absolute temperature, °K

C & n = Constants, dependent on crop type and temperature.

Substitute the values in equation

$$1\text{-}0.1=e^{-CTM_e^n}$$

$$\ln 0.9 = -6.12\times10^{-5}\times(60+273)\times M_e^{2.31}$$

$$-0.10536 = -0.0204\times M_e^{2.31}$$

$$M_e^{2.31} = 5.165 = M_e = 5.165^{\frac{1}{2.31}}$$

$M_e = 2.046\%$ db

Problem 5

Determine the values of c and n from the Henderson's equation for the following data obtained from thin layer paddy drying studies:

RH = 50%, t = 60°C, $M_e = 22\%$

RH = 70%, t = 60°C, $M_e = 35\%$

Henderson's equation is expressed as,

$$1\text{-RH} = \exp[-CTM_n^e]$$

Substituting conditions (i) in Henderson's equation we get

$1 - 0.5 = \exp[- C \times (60+273) \times 22^n]$

$0.5 = \exp[- C \times 333 \times 22^n]$

$\ln (0.5) = - C \times 333 \times 22^n$

$- 0.693 = - C \times 333 \times 22^n$ ....(i)

Substituting conditions (ii) in Henderson's equation we get

$1 - 0.7 = \exp[- C \times (60+273) \times 35^n]$

$0.3 = \exp[- C \times 333 \times 35^n]$

$\ln (0.3) = - C \times 333 \times 35^n$

$- 1.204 = - C \times 333 \times 35^n$ ....(ii)

Dividing equation (ii) by equation (i),

$$\frac{-1.204}{-0.693} = \frac{-C\times333\times35^n}{-C\times333\times22^n}$$

$$1.737 = \left(\frac{35}{22}\right)^n$$

$1.737 = (1.591)^n$

$\log(1.737) = \log(1.591)^n$

$\log(1.737) = n \times \log(1.591)$

$$0.2398 = n \times 0.2017$$

$$n = \frac{0.2398}{0.2017} = 1.19$$

Substituting the value of n in equation (i),

$$-0.693 = -C \times 333 \times (22)^{1.19}$$

$$0.693 = C \times 333 \times 39.581$$

$$C = \frac{0.693}{333 \times 39.581} = 5.256 \times 10^{-5}$$

Exercises

1. Calculate the amount of moisture evaporated from 1500 kg of grain for drying it from initial moisture content of 25 % to final moisture content of 13 % on wet basis.
2. Five tonnes of parboiled paddy is to be dried from 35 % to 13 % moisture content on wet basis. Calculate the amount of moisture to be evaporated and dried products.

17

Determination of Drying Rate of Agricultural Products

Aim

- To study the determination of drying rate of agricultural products

Problem

In a drying experiment on paddy at an air temperature of 60°C, the following data were obtained. Prepare a drying rate curve for the experiment. Initial weight of the sample is 1250 gm and initial moisture content is 35% (d.b).

S.No	Drying time in min	Moisture removed in gm
1.	0	0.0
2.	30	45.3
3.	60	64.7
4.	90	86.4
5.	120	107.5
6.	150	125.8
7.	180	151.4
8.	210	180.6
9.	240	224.8
10.	270	267.5

Solution

Initial moisture content = 35% (db)

Initial weight of the sample = 1250 gm

In general rule, the amount of bone dry material in the sample should be determined through the wet basis moisture content

Relationship between dry basis and wet basis moisture content:

$$MC\,(db)\,\% = \frac{MC\,(wb)}{100 - MC\,(wb)} \text{ x } 100 \qquad \text{or}$$

$$MC\,(wb)\,\% = \frac{MC\,(db)}{100 + MC\,(db)} \text{ x } 100$$

$$MC\,(wb)\,\% = \frac{35}{100 + 35} \text{ x } 100 = 25.926\%$$

Amount of bone dry material in the sample

$$= 1250 - \left(1250 \text{ x } \frac{25.926}{100}\right) = 925.926 \text{ gm}$$

OR

$$= \frac{100}{135} \text{ x } 1250 = 925.926 \text{ gm}$$

Amount of moisture present in the sample = 1250 - 925.926 = 324.074 gm

After 30 minutes,

Amount of water removed from the sample = 45.3 gm

Moisture present in the sample = 324.074 - 45.3 = 278.774 gm

Moisture content of the sample

$$= \frac{324.074 - 45.3}{925.926} \text{ x } 100 = 30.11\%\,(db)$$

Drying rate (R)

Drying rate in gram of water per minute per 100 gram of bone dry material

$$R = \frac{\text{Amount of moisture removed}}{\text{Time taken} \times \left(\dfrac{\text{Total bone dry weight of sample in gm}}{100}\right)}$$

$$R = \frac{45.3}{30 \times \left(\dfrac{925.926}{100}\right)} = 0.163$$

After 60 minutes,

Amount of water removed from the sample = 64.7 gm

Moisture present in the sample = 324.074 - 64.7 = 259.374 gm

Moisture content of the sample

$$MC = \frac{324.074 - 64.7}{925.926} \times 100 = 28.01\% \text{ (db)}$$

$$R = \frac{64.7}{60 \times \left(\frac{925.926}{100}\right)} = 0.1165$$

After 90 minutes,

Amount of water removed from the sample = 86.4 gm

Moisture present in the sample = 324.074 - 86.4 = 237.674 gm

Moisture content of the sample

$$MC = \frac{324.074 - 86.4}{925.926} \times 100 = 25.67\% \text{ (db)}$$

$$R = \frac{86.4}{90 \times \left(\frac{925.926}{100}\right)} = 0.1037$$

After 120 minutes,

Amount of water removed from the sample = 107.5 gm

Moisture present in the sample = 324.074 - 107.5 = 216.574 gm

Moisture content of the sample

$$MC = \frac{324.074 - 107.5}{925.926} \times 100 = 23.39\% \text{ (db)}$$

$$R = \frac{107.5}{120 \times \left(\frac{925.926}{100}\right)} = 0.0967$$

After 150 minutes,

Amount of water removed from the sample = 125.8 gm

Moisture present in the sample = 324.074 - 125.8 = 198.274 gm

Moisture content of the sample

$$MC = \frac{324.074 - 125.8}{925.926} \times 100 = 21.42\,\%\,(db)$$

$$R = \frac{125.8}{150 \times \left(\frac{925.926}{100}\right)} = 0.0906$$

After 180 minutes,

Amount of water removed from the sample = 151.4 gm

Moisture present in the sample = 324.074 - 151.4 = 172.674 gm

Moisture content of the sample

$$MC = \frac{324.074 - 151.4}{925.926} \times 100 = 18.65\,\%\,(db)$$

$$R = \frac{151.4}{180 \times \left(\frac{925.926}{100}\right)} = 0.0908$$

After 210 minutes,

Amount of water removed from the sample = 180.6 gm

Moisture present in the sample = 324.074 - 180.6 = 143.474 gm

Moisture content of the sample

$$MC = \frac{324.074 - 180.6}{925.926} \times 100 = 15.5\,\%\,(db)$$

$$R = \frac{180.6}{210 \times \left(\frac{925.926}{100}\right)} = 0.0929$$

After 240 minutes,

Amount of water removed from the sample = 224.8 gm

Moisture present in the sample = 324.074 - 224.8 = 99.274 gm

Moisture content of the sample

$$MC = \frac{324.074 - 224.8}{925.926} \times 100 = 10.72\,\%\,(db)$$

$$R=\frac{224.8}{240\times\left(\frac{925.926}{100}\right)}=0.1012$$

After 270 minutes,

Amount of water removed from the sample = 267.5 gm

Moisture present in the sample = 324.074 - 267.5 = 56.574 gm

Moisture content of the sample

$$MC=\frac{324.074-267.5}{925.926}\times 100=6.11\%\,(db)$$

Sl. No	Drying time (min)	Moisture removed (g)	Moisture present in the sample (g)	Moisture content (db) %	Drying rate (R) (g of water/min/100 g of bone dry material)
1	0	0.0	324.074	35.00	
2	30	45.3	278.774	30.11	0.1630
3	60	64.7	259.374	28.01	0.1165
4	90	86.4	237.674	25.67	0.1037
5	120	107.5	216.574	23.39	0.0967
6	150	125.8	198.274	21.42	0.0906
7	180	151.4	172.674	18.65	0.0908
8	210	180.6	143.474	15.5	0.0929
9	240	224.8	99.274	10.72	0.1012
10	270	267.5	56.574	6.11	0.1070

$$R=\frac{267.5}{270\times\left(\frac{925.926}{100}\right)}=0.107$$

Result

Drying rare curve

Drying rate curve is obtained by plotting the rate of drying against the moisture content.

18

Study of Drying Characteristics of Grains and Evolution of Lab Model Thin Layer Dryer

Aim

- To draw the drying characteristic curve for given grain by thin layer drying.
- To determine the drying constant, heat utilization factor and coefficient of performance.

Materials required

- Lab model thin layer dryer
- Thermometer
- Hot-air oven
- Weighing balance
- Grains

Description

The lab model thin layer dryer (Fig.18.1) consists of a blower, heater assembly, plenum chamber and holding bin. The blower used is a built in type run by single-phase electric supply and has provision to adjust the airflow rate at the suction side. The delivery end is connected to the heater assembly.

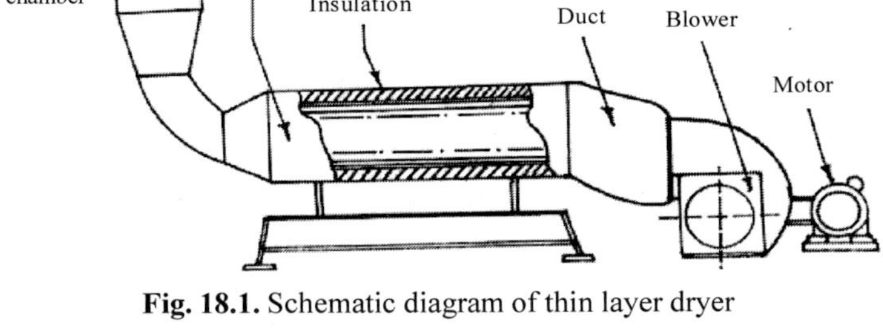

Fig. 18.1. Schematic diagram of thin layer dryer

The heater assembly is made in cylindrical shape and 3 numbers of heating coils of capacity each 1000 w has been provided and connected in series. The other end of the heater assembly is connected to plenum chamber.

At the inlet of the heater assembly an orifice is provided to measure the airflow rate with the help an u-tube monometer in terms of velocity head. In the plenum chamber, a container is to hold the grain to be dried. This container can be removed at periodical intervals for weighing the grain along the bin to avoid spilling of grain during handling/transferring. A stem type thermometer is provided in the 'plenum chamber' and is connected in the circuit to control the temperature of hot air. A timer switch is also provided in the circuit to put off the complete equipment at desired time interval of drying.

Procedure

- The empty weight of the grain holding container and weight of sample taken for drying is noted.
- To determine initial moisture content minimum three samples each 10 grams are taken and kept in the oven at 130±2 f c for one hour.
- The container with grains is placed in the plenum chamber and drying is allowed to take place at desired airflow rate and temperature of hot air by keeping the timer switch at desired time intervals of 10 minutes.
- For every 10 minutes interval temperature of air at inlet (suction), hot air and exhaust air from container are recorded.
- The change in weight at every 10 minutes is also recorded, until a constant value is reached.

Observations

Time of drying process	t	=	h
Weight of sample (grain) taken for drying	W_i	=	kg
Final weight of the sample	W_f	=	kg
Weight of the dry matter of the sample	W_d	=	kg
Weight of the sample at time 't'	W_t	=	kg
Weight of moisture removed from the sample	W_m	=	kg
Temperature of ambient air	T_1	=	°C
Temperature of heated air used for drying	T_2	=	°C
Temperature of exhaust air after drying	T_3	=	°C

Calculations

Initial moisture content of the sample (hot-air oven method):

$$M_i, \%(wb) = \frac{\text{Initial weight} - \text{Final weight}}{\text{Initial weight}} \times 100$$

Weight of initial moisture content in the sample (X_i) = $W_i \times M_i$

Weight of initial bone-dry matter in the sample (W_d) = W_i - X_i

Moisture content at any time, % (db):

$$M_i, \%(db) = \frac{W_t - W_d}{W_d} \times 100$$

Initial moisture content, % (db):

$$M_i, \%(db) = \frac{M_i, \%(wb)}{1 - M_i, \%(wb)} \times 100$$

Weight of moisture removed from the sample during drying (W_m):

$$W_m = W_d \times [M_i, \%\ db - M_t, \%\ db]$$

From Newton's law of cooling or heating

$$\frac{M_t - M_e}{M_i - M_e} = e -^{kt}$$

$$M_e = \frac{W_{eg} - W_d}{W_d}$$

Where,

M_t = moisture content at any time, % (db)

M_i = initial moisture content, % (db)

M_e = equilibrium moisture content, % (db)

W_{eg} = weight of sample at equilibrium, g

W_d = weight of dry material, g

t = drying time, h

K = drying constant

i) Moisture ratio (MR)

$$MR = \frac{M_t - M_e}{M_i - M_e}$$

ii) Drying constant (K)

$$K = \frac{1}{t} \ln \frac{M_t - M_e}{M_i - M_e}$$

iii) Rate of drying

Rate of drying may be defined as the amount of water removed from the unit weight of food material per unit time. The rate of drying, R can be expressed as follows:

$$R = \frac{\text{Amount of moisture removed, g}}{\text{Time Taken} \times \left(\frac{\text{Total weight of bone dry material, g}}{100} \right)}$$

Dryer performance

i) **Heat utilization factor (H.U.F):** H.U.F defined as the ratio of drop in temperature of air during drying due to cooling to the increase in temperature of ambient air due to heating.

$$H.U.F = \frac{\text{Drop in temperature of drying air, °C}}{\text{Increase in temperature of ambient air, °C}} = \frac{T_2 - T_3}{T_2 - T_1}$$

ii) **Coefficient of performance (C.O.P):**

$$C.O.P = \frac{T_2 - T_3}{T_2 - T_1}$$

iii) **Effective heat efficiency (E.H.E):** Sensible heat of heated air should be considered for estimation of effective heat efficiency of the drying system. The E.H.E is mathematically defined as follows:

$$E.H.E = \frac{T_2 - T_3}{T_2 - T_w}$$

Where,

T_w = wet bulb temperature of the drying air, °C

iv) **Thermal efficiency ():** Thermal efficiency of the drying system can be estimated using following relationship:

$$\eta = \frac{\text{Amount of heat utilized}}{\text{Amount of heat supplied}} = \frac{T_2 - T_3}{T_2 - T_1}$$

v) Overall thermal efficiency (η_0):

$$\eta_0 = \frac{\gamma \times \delta}{F \times \lambda} \times 100$$

Where,

η_0 = overall thermal efficiency,

γ = amount of water vapour removed, kg

δ = latent heat of evaporation, kcal/kg

F = amount of fuel, kg

λ = calorific value of fuel, kcal/kg

vi) **Drying characteristics of the sample is drawn by plotting**

a) Time versus Moisture content

b) Time versus Drying rate

c) Time versus Moisture ratio

Results

i) Drying rate (R) =

ii) Moisture ratio (MR) =

iii) Drying constant (K) =

iv) Heat utilization factor (HUF) =

v) Coefficient of performance of the dryer (COP) =

vi) Effective heat efficiency (EHE) =

vii) Thermal efficiency (η) =

19

Study of Drying Characteristics of The Grains by Fluidized Bed Drying

Aim

- To draw the drying characteristics curve for the given grains by fluidized bed drying.

Materials required

- Lab model fluidized bed drier
- Weighing balance
- Hot-air oven
- Grains

Theory

Fluidized bed drying is a method of drying by keeping the material in fluidized/ floating condition inside a chamber by hot air. In this drying process higher rate of moisture migration takes place.

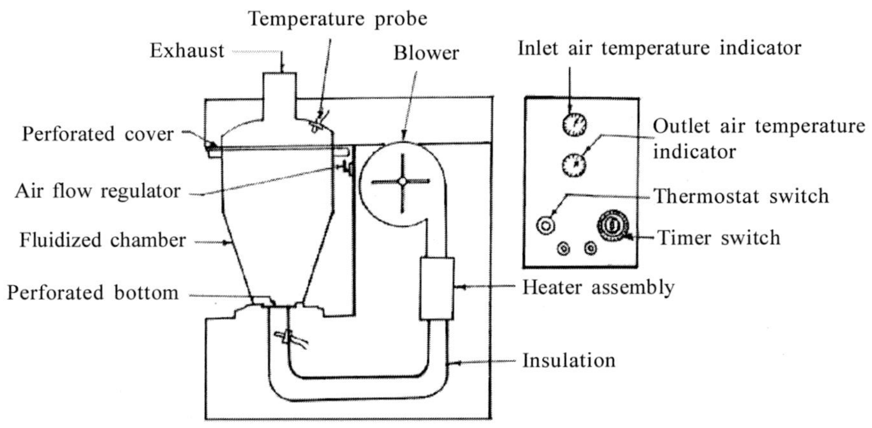

Fig. 19.1. Schematic diagram of lab model fluidized bed dryer

This method is normally used for the materials which have higher initial moisture content and at the same time requires to be dried quickly such as vegetable seeds. The drying characteristic curve indicate the relationship between the moisture content against time drying rate against time.

Description

The lab model fluidized bed dryer (Fig.19.1) consists of a built in type blower run by a motor. This blower in takes the atmospheric air and the butterfly valve has been provided to regulate the entry of the air. In the outlet of the blower, a heater is provided to heat the air for drying purpose. The temperature of the air is controlled by the thermostat switch connected in the heater circuit. The hot air enters into the truncated cone shaped fluidizing chamber from the bottom. A wire mesh has been fixed at the bottom to hold the material to be dried in the chamber. A removal white mesh is provided at the top of the chamber to avoid the material carried away along the air. There are two temperature sensors in the inlet and outlet of the hot air. A timer switch of range 0 to 60 minutes is included in the circuit to put off the whole unit at desired set time. The top of unit is provided with a hinged door for easy mounting and removal of the fluidizing chamber.

Procedure

To determine the initial moisture content minimum three samples, each 10 grams are taken and kept in the oven at 130 ± 2°C for one hour.

- The empty weight of drying chamber and weight of sample taken for drying is noted. The drying chamber with grains is placed in position in the lab model drier.
- The airflow rate to the drying chamber is adjusted by controlling the butterfly valve provided. By this adjustment the given grain to be dried is kept in fluidized condition.
- This regulation of air is kept at constant throughout the experiment and the temperature of air is kept at desired level by adjusting the thermostat and the value is noted.
- The timer switch is kept for 10 minutes operation, by which the heating will be stopped automatically.
- The change is weight of sample is noted for every 10 minutes , until a constant weight is reached.

Observations

Time of drying process	t	=	h
Initial weight of sample (grain)	W_i	=	kg
Final weight of the sample	W_f	=	kg
Weight of the dry matter of the sample	W_d	=	kg
Weight of the sample at time 't'	W_t	=	kg
Weight of moisture removed from the sample	W_m	=	kg
Initial temperature of the drying air	T_1	=	°C
Temperature of hot air	T_2	=	°C
Temperature of exhaust air	T_3	=	°C

Calculations

Initial moisture content of the sample (hot-air oven method):

$$M_i, \% \text{ (wb)} = \frac{\text{Initial weight-Final weight}}{\text{Initial weight}} \times 100$$

Weight of initial moisture content in the sample (X_i) = $W_i \times M_i$

Weight of initial bone-dry matter in the sample (W_d) = W_i - X_i

Moisture content at any time, % (db):

$$M_i, \% \text{ (db)} = \frac{W_t - W_d}{W_d} \times 100$$

Initial moisture content, % (db):

$$M_i, \% \text{ (db)} = \frac{M_i, \% \text{ (wb)}}{1 - M_i, \% \text{ (wb)}} \times 100$$

Weight of moisture removed from the sample during drying (W_m):

$$W_m = W_d \times [M_i, \% \text{ db} - M_t, \% \text{ db}]$$

i) Drying rate (R)

$$R = \frac{W_m}{t \times \frac{W_d}{100}}$$

ii) Moisture ratio (MR)

$$MR = \frac{M_t - M_e}{M_i - M_e}$$

iii) Drying constant (K)

$$K = \frac{1}{t} \ln \frac{M_t - M_e}{M_i - M_e}$$

iv) Heat utilization factor (H.U.F)

$$H.U.F = \frac{T_2 - T_3}{T_2 - T_1}$$

v) Coefficient of performance (C.O.P)

$$C.O.P = \frac{T_3 - T_1}{T_2 - T_1}$$

vi) Effective heat efficiency (E.H.E)

$$E.H.E = \frac{T_2 - T_3}{T_2 - T_w}$$

vii) Thermal efficiency (η)

$$\eta = \frac{T_2 - T_3}{T_2 - T_1}$$

Results

i) Drying rate (R) =

ii) Moisture ratio (MR) =

iii) Drying constant (K) =

iv) Heat utilization factor (HUF) =

v) Coefficient of performance of the dryer (COP) =

vi) Effective heat efficiency (EHE) =

vii) Thermal efficiency (η) =

20

Study on Performance Evaluation of Seed Cleaner Cum Grader

Aim

- To study the working principle and cleaning efficiency of air-screen seed cleaner cum grader

Apparatus required

- Lab model air-screen seed cleaner cum grader
- Seed sample
- Weighing balance

Construction

The air-screen seed cleaner uses three cleaning systems

i) Blowing or aspiration

ii) Scalping at top screens

iii) Grading at lower screens

The screening unit is composed of double or multiple screens. These screens are tightened together and suspended by hangers in such a manner that these have horizontal oscillating motion and slightly vertical motion. These two motions in combination move the grain down the screen and at the same time toss sufficiently above the screen so that the bed of grain is properly stirred.

The slope of the screen is adjustable to control the rate of downward travel of the grain. The screens are available in various shapes like; round, slotted holes, oblong and triangular. Sometimes the holes of the screen are clogged when fine degree of sorting is made by the machine. To avoid the clogging, the screens are generally fitted with a brush which moves under the screen and pushes the clogged material back through the screen (Fig.20.1).

A blower is provided at the bottom which is operated by 0.25 hp single electric motor and with crank mechanism, by which the sieve box reciprocates. The power is transmits through belt drive mechanism. A stream of air generated from a blower is used to separate the particles lighter than the product. Effective cleaning or winnowing of grains is achieved by regulating the airflow rate. For this purpose, a hand operated air regulator is provided. Guides are provided in the sieve box to insert two numbers of sieves for cleaning cum grading.

Working principle

A simple vibratory type air-screen cleaner is shown in Fig. 20.2. It consists of a feed hopper, blower, sieve box and grain outlets. Based on the grains to be cleaned and graded, the sieves can be selected and inserted into guides provided in the sieve box. Top screen should be scalping screen and bottom screen for grading. The feed hopper is placed above the sieve box. The grain passes from the feed-hopper reaches directly to the upper screen. There are outlets to collect overflow and underflow from the top and bottom sieves respectively. Coarse impurities such as stones, straw particles, etc are screened off by upper screen and discharged out through an outlet. The grain falls through on to the lower screen, where the sand and the dust particles are screened off. The overflow of bottom sieve is subjected to the stream of air provided from the blower and the chaff and other impurities are removed. The cleaned grains are collected at the bottom and the impurities and chaff are blown out.

Fig.20.1. Lab model air-screen cleaner cum grader

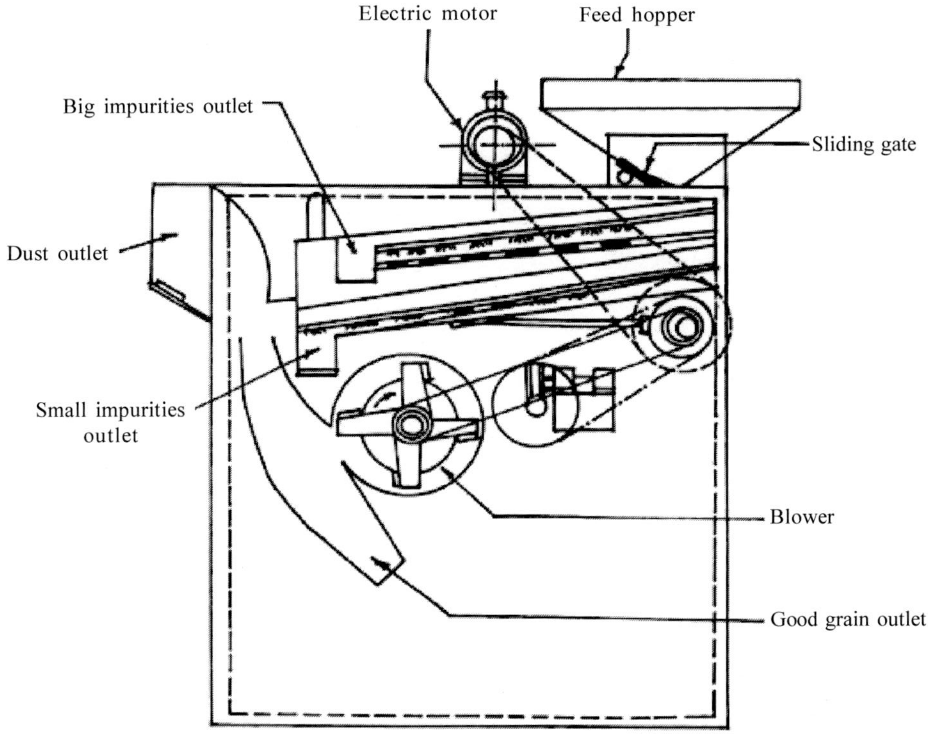

Fig.20.2. Schematic diagram of seed cleaner cum grader

Test Procedure

- About one kilogram of grain (Paddy) mixed with impurities (chaff, stones) was taken for experiment.
- Minimum three samples of each 50 grams were taken from the feed and good grains were separated manually.
- Average value of fraction of good grain in the feed was calculated, Grains were fed in the feed hopper and the machine was operated by adjusting airflow rate accordingly.
- Time taken for cleaning and grading was noted. The good grain and chaff, stones were collected from respective outlets.
- Minimum three samples (50g) were taken from these outlets and good grains were separated manually.
- The average values of fraction of good grain in good grain outlet, chaff outlet, stone outlet were calculated.

Observation and calculation

Weight of grain taken	(W)	=	kg
Time taken for cleaning and grading	(T)	=	hr
Fraction of good grain in feed	(F)	=	
Fraction of good grain in grain outlet	(E)	=	
Fraction of good grain in impurities outlet	(G)	=	
Capacity of cleaner cum grader =	W/T	=	kg/h

Effectiveness of cleaner cum grader

$$\text{Fraction} = \frac{\text{weight of material-weight of ompurities}}{\text{weight of material}}$$

$$\eta = \frac{E \times (F-G) \times (E-F) \times (1-G)}{F \times (E-G)^2 \times (1-F)} \times 100$$

Result

Capacity of cleaner cum grader =

Effectiveness of cleaner cum grader =

21

Performance Evaluation of Inclined Belt Separator

Aim

- To determine the effectiveness of inclined belt separator
- To determine the capacity of inclined belt separator

Materials required

- Inclined belt separator
- Weighing balance
- Feed material
- Sample divider

Theory

The separation by inclined belt draper takes place due to difference in shape and surface texture of the material. The frictional properties can be utilized for the separation of a mixture of grains of almost same size. The mixture to be separated is fed over the centre of an inclined draper belt moving in upward direction. The round and smooth grains roll or slide down the draper at faster rate than the upward motion of the belt, and these are discharged in a hopper. The flat shape or rough surfaced particles are carried to the top of the inclined draper and dropped off into another hopper. The effectiveness of inclined belt separator is calculated by:

$$\text{Fraction} = \frac{\text{weight of material-weight of impurities}}{\text{weight of material}}$$

$$\eta = \frac{E \times (F\text{-}G) \times (E\text{-}F) \times (1\text{-}G)}{F \times (E\text{-}G)^2 \times (1\text{-}F)} \times 100$$

Where,

η = Effectiveness of inclined belt separator, %

F = Fraction of clean seed in feed

E = Fraction of clean seed at clean seed outlet

G = Fraction of clean seed in foreign matter outlets

Description

The unit consists of a belt wrapped over two rollers. the top roller is mounted in slots to adjust the slope of the belt (Fig.21.1). The idle roller is provided in between the two rollers for giving tension to the belt. The hopper is designed to hold 5 kg of seeds. A regulator is provided at the bottom of the feed hopper for adjusting the feed rate. A 'U' shaped guide is kept beneath the belt to the entire length to avoid belt moving aside. Two rubber sheets are provided on either side of the wall to prevent the grains from getting inside the belt and the roller.

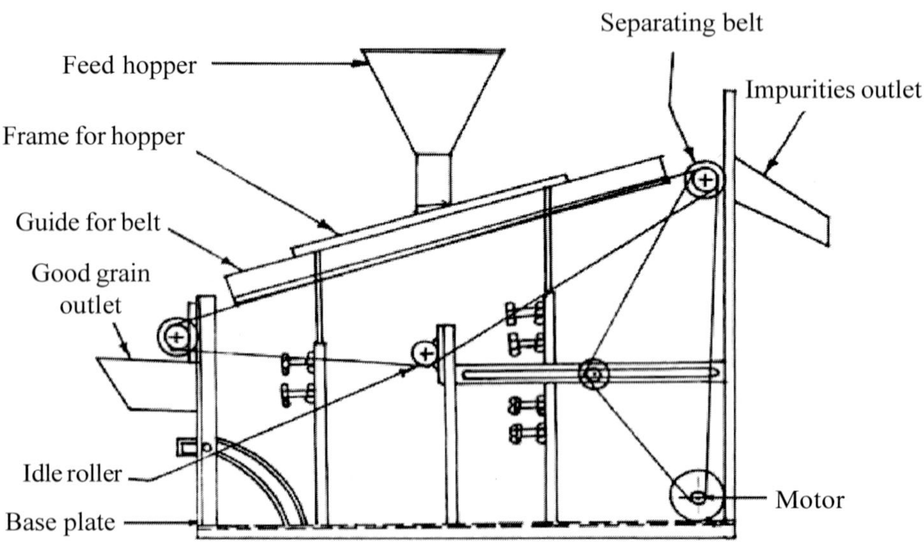

Fig. 21.1. Schematic diagram of inclined belt separator

The belts of different degrees of toughness may be used as a draper for separate materials. If rolling tendencies of the grain are predominant, the rough canvas belt may be used. The smooth, plastic belt may be used in case sliding action is desired for the lower fraction. Feed rate, speed of draper and angle of inclination are other important variables for effective separation of dissimilar materials.

Test procedure

- About one kg of feed is taken and filled in the feed hopper.
- Three samples about 50 grams each are taken and the good grains in the mixture is separated manually using sample divider.
- The fraction of good grains present in the feed is calculated from the samples.
- The belt separator is switched on and the mixture is fed uniformly in the belt.
- Time taken for separation is noted.
- The fraction of good grain is calculated separately from the samples collected at good grain outlet and impurity outlets.

Observations

$$\text{Fraction}=\frac{\text{weight of material-weight of impurities}}{\text{weight of material}}$$

Weight of feed taken	(W)	=	kg
Time taken for separation	(T)	=	h
Fraction of clean seed in feed	(F)	=	
Fraction of clean seed at clean seed outlet	(E)	=	
Fraction of clean seed in foreign matter outlets	(G)	=	

Calculations

Capacity of inclined belt separator = W/T = kg/h

Effectiveness of inclined belt separator, %

$$\eta=\frac{E\times(F-G)\times(E-F)\times(1-G)}{F\times(E-G)^2\times(1-F)}\times100$$

Result

i) Capacity of inclined belt separator___________kg/h

ii) Effectiveness of inclined belt separator________%

22

Performance Evaluation of Specific Gravity Separator

Aim

- To evaluate the performance of the gravity separator.

Materials required

- Specific gravity separator
- Weighing balance
- Sample divider
- Stopwatch
- Grains

Theory

The specific gravity separator also known as stoner separator separates the grain mass into two fractions as per the difference in specific gravity. The main purpose of this machine is to remove dust and stone from the grain mass effectively. The grain with impurities is distributed uniformly by dispersing plate. The flow of the mixture is controlled by a feed control valve. The vibrating deck provided with corrugated sieve mounted with required inclination makes the stone to move upward and grains downward to effect the separation. Light and small impurities in the grain are aspirated by an aspirator. The immature grain is separated in the expansion chamber. The efficiency of specific gravity separator is calculated by:

$$\text{Fraction} = \frac{\text{weight of material-weight of impurities}}{\text{weight of material}}$$

$$\eta = \frac{E \times (F - G) \times (E - F) \times (1 - G)}{F \times (E - G)^2 \times (1 - F)} \times 100$$

Where,

h = Effectiveness of inclined belt separator, %

F = Fraction of clean seed in feed

E = Fraction of clean seed at clean seed outlet

G = Fraction of clean seed in foreign matter outlets

Description

The machine consists of feed hopper, vibrating deck, aspirator, stone and grain outlets, regulator, inclined reciprocating tray having convex slots all over surface and blower at bottom. The blower is operated by 2 HP single phase electric motor and completely enclosed in an airtight compartment. The air from the blower is passed through a flexible duct to the vibrating duct which powered by the same motor through an eccentric connecting rod. The deck is mounted on the base with suitable bearings and crank. In vibrating deck a corrugated sieve with upward openings for air delivery is provided and the inclination of the deck can be adjusted. The feed hopper is mounted above this deck and feed regulator is provided to adjust the inflow. The stone outlet has a regulator to control the size of the stones being separated. the air flow rate at the suction side can be adjusted (Fig.22.1).

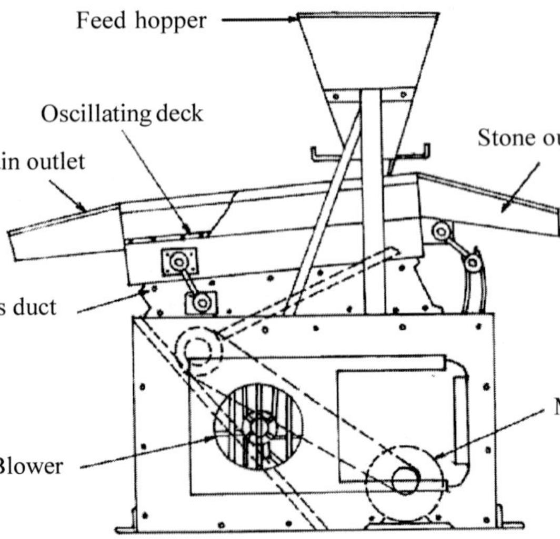

Fig. 22.1. Schematic diagram of specific gravity separator

When the machine is operated, the mixture of grain and stone is fed at the top of the reciprocating ray, the stones having higher specific gravity slowly go down and occupy the bottom layer of the mixture and thus come in contact with the reciprocating tray which operated by oscillating mechanism. A large amount of air is blown from underneath through the slotted separating tray thus the grain bed is suspended in the air stream.

The materials while the reciprocating action of the tray separates the heavy stone particles from the light grain particles. The heavier stones are carried to the top of the tray by its movement. Due to the difference in specific gravity and vibration, the stones will move upward and discharged through the stone outlet.

Since the direction of movement is more inclined than the inclination of the tray, the heavier particles in contact with the reciprocating tray are lifted and are move towards the top end of the tray. The lighter particle floats on the stone and moves downward by gravity without any middling product. The strong air current through the slots of the tray, directed towards the movement of the grain further facilitates in lifting and shifting them downwards. The separation can be controlled by adjusting the feed rate, slope of the deck, deck vibration and the air flow rate.

Procedure

- A known weight of good grain mixed with few kg of stone in it is taken as the raw material for the experiment.
- Using sample divider three samples of about 50 g each are taken.
- Grains and stones are manually separated and weighed.
- The fraction of good grains present in the feed is calculated from the samples.
- Then raw materials is fed into the unit and airflow rate and feed regulator are adjusted.
- Time taken for separation is noted. Samples are collected from good grain outlet and stone outlet.
- From these good grains are separated and fraction of good grain in these outlets are calculated.

Observations

$$\text{Fraction} = \frac{\text{weight of material-weight of impurities}}{\text{weight of material}}$$

Weight of feed taken	(W)	=	kg
Time taken for separation	(T)	=	h
Fraction of clean seed in feed	(F)	=	
Fraction of clean seed at clean seed outlet	(E)	=	
Fraction of clean seed in foreign matter outlets	(G)	=	

Calculations

Capacity of destoner = W/T = kg/h

Efficiency of destoner, %

$$\eta = \frac{E \times (F - G) \times (E - F) \times (1 - G)}{F \times (E - G)^2 \times (1 - F)} \times 100$$

Result

i) Capacity of specific gravity separator _____________ kg/h.

ii) Efficiency of specific gravity separator _____________ %.

23

Performance Evaluation of Spiral Separator

Aim

- To determine the effectiveness of the spiral separator.

Materials required

- Spiral separator
- Weighing balance
- Sample divider
- Stopwatch
- Grain mixture (round and others)

Theory

The spiral separator separates the grains as per their roundness. The main component of the separator is a stationary, open screw conveyor standing on one end. The mixture is fed at the top of the unit. The round materials of the mixture pick up speed as they slide or roll down the inclined surface until their centrifugal force become sufficient enough to throw them in the outer helix. While the non-round materials are caught in the inner helix and are discharged through a separate spout.

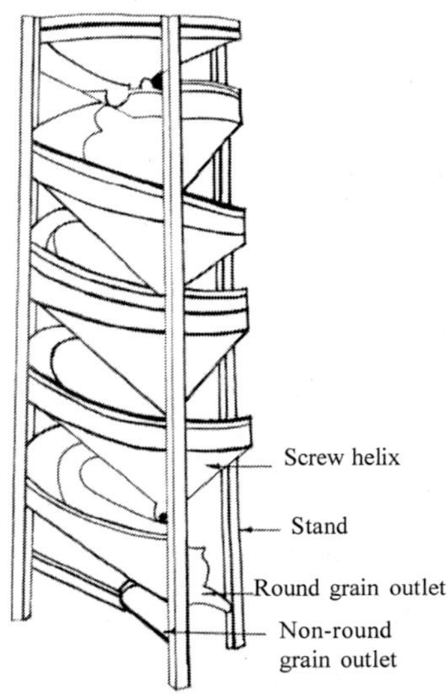

Fig. 23.1. Spiral separator

There is no moving part in the spiral separator. The rate of feeding is the only adjustable component. The feeding should be such that each grain/particle rolls independently for effective separation. The main limitation of the spiral is lack of flexibility. Separation of mustard, rape, soybean, wild, peas or other round seeds can be performed from wheat, flax, oats etc. This device is less versatile as compared to other mechanical cleaners, but it is simple, inexpensive and quite useful for seed cleaning purposes.

Procedure

- Take about one kilogram of seed mixture contain round and oblong size and is filled in the feed hopper.
- Three sub samples of about 50 g each are taken from the sample using sample divider.
- The good grains in these samples should be separated manually.
- The material should be allowed to fall from the feed hopper gradually.
- Time for separation should be noted.
- The fraction of good grains should be collected separately from samples the samples at good grain outlet and impurity outlets.

Observations

$$\text{Fraction} = \frac{\text{weight of material-weight of impurities}}{\text{weight of material}}$$

Weight of feed taken	(W)	=	kg
Time taken for separation	(T)	=	h
Fraction of clean seed in feed	(F)	=	
Fraction of clean seed at clean seed outlet	(E)	=	
Fraction of clean seed in foreign matter outlets	(G)	=	

Calculations

Capacity of spiral separator = W/T = kg/h

Effectiveness of spiral separator, %

$$\eta = \frac{E\,(F - G)\,(E - F)\,(1 - G)}{F\,(E - G)^2\,(1 - F)} \times 100$$

Result

i) Capacity of spiral separator __________ kg/h.

ii) Efficiency of spiral separator __________ %.

24

Performance Evaluation of Groundnut Decorticator

Aim

- To determine the efficiency of groundnut decorticator

Apparatus required

- Manually operated groundnut decorticator
- Weighing balance
- Groundnut pods

Theory

The removal of kernels from the groundnut pods by hand picking is laborious, painful, time consuming and costly process. The manually operated groundnut decorticator is most suitable for small growers. As compared to hand picking, it is economical, saves time and reduces drudgery of workers. The performance of the decorticator is expressed in terms of output/ capacity, percentage of un-decorticated pods and percentage of breakage (Fig.24.1). The above terms are defined as:

Let,

C = capacity of decorticator, (kg/h)

Q = quantity of pod decorticated, kg

Q_1 = quantity of un-decorticated pods, kg

T = time taken for decorticating of kg of pods, h

η= decortications efficiency, percent

B = breakage of kernels, percent

N_1 = unbroken kernels in each sample, kg

N_2 = broken kernels in each sample, kg

Capacity of decorticator, kg/h = $\frac{Q}{T}$

$$\eta = \frac{Q\text{-}Q_1}{Q} \times 100$$

$$B = \frac{N_2}{N_1\text{-}N_2} \times 100$$

Description

The decorticating machine has to perform the following functions:

i) Feeding of groundnuts to shelling unit

ii) Shelling of groundnuts and separating nuts from the shell

iii) Cleaning the nut from the mixture.

The feeder mechanism consists of hopper, feeder shaft, and pawl and ratchet wheel to drive the shaft. Groundnuts are fed to the machine at a uniform rate. Arrangements are also made to vary the rate of feeding by adjusting the stroke of powl. The shelling unit is provided with two main parts, namely crushing plates and grate. Crushing plates are made of close-grained semi steel. Six small channel grooves are mounted on radial arms of the main rotor shaft which rotates at about 100 rpm. The main rotor shaft is operated, the external grooved surfaces of the channels are shell the groundnuts. The crushing plates are mounted at an angle resulting in unequal gaps on both ends between the plates and the grate (Fig.24.2). The minimum uniform gap of 25 to 32 cm between the lower edge of the plates and inner surface of the grate gives satisfactory operation.

The fan, which is operated at 450 rpm by the main rotor shaft, is the basic part of the cleaning section. The blades of the fan are made of iron sheets bolted to angle iron arms on both sides. The blast of air created by the fan is strong enough to blow away the broken empty shells so that the heavy nuts fall into the bottom pan by gravity. The hand operated machine works on the same principle as the power operated machine and are operated by two persons although it can also be run by a 1.5 hp engine. The capacity of the machine is about 175 to 200 kg of groundnut per hour when operated by hand and with mechanical power it will shell about 500 kg groundnuts in an hour.

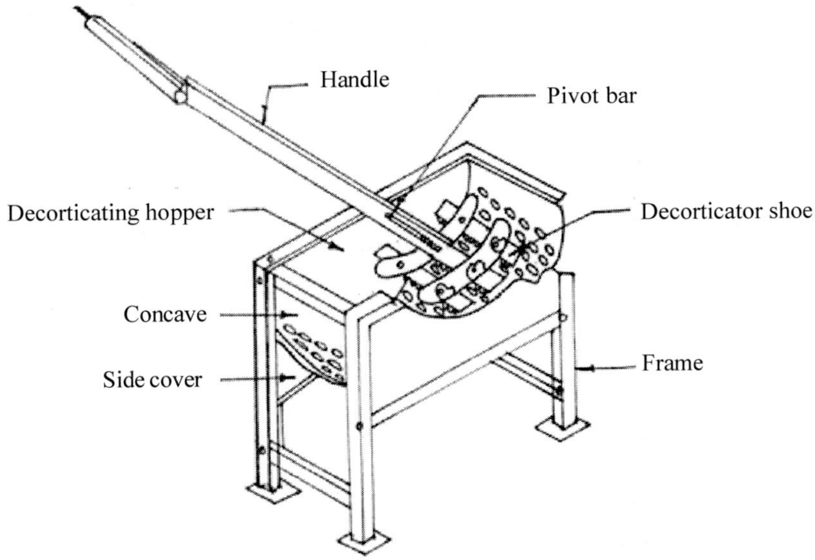

Fig.24.1. Manually operated groundnut decorticator

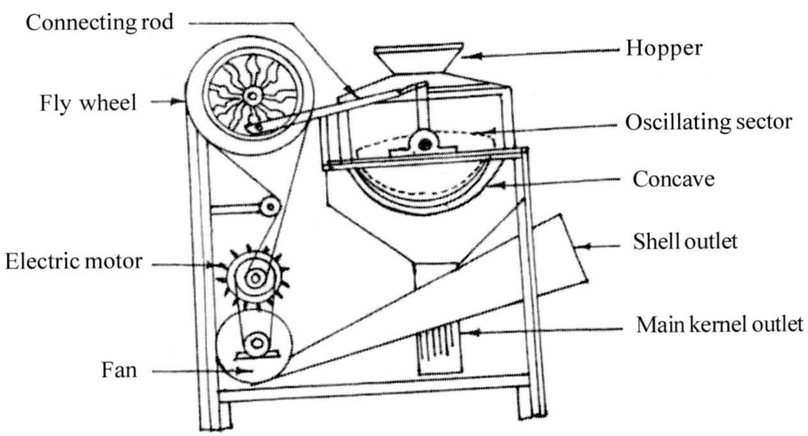

Fig.24.2. Power operated groundnut decorticator

Procedure

- The clearance between the concave and oscillating sector is adjusted according to the size of the groundnut pods.
- One kg of groundnut is weighed and fed into the unit. The pods are shelled by operating the oscillating sector.

- Time taken for shelling of pods is noted. The kernels, husk and unshelled pods are collected at bottom.
- From the product, kernels, broken and unshelled pods are separated and weighed.

Observation

Quantity of pod decorticated	(Q)	=	kg
Time taken for shelling	(t)	=	h
Weight of un-decorticated pods	(Q_1)	=	kg
Weight of broken kernels	(N_1)	=	kg
Weight of unbroken kernels	(N_2)	=	kg

Calculation

$$\text{Capacity of decorticator} = \frac{Q}{T} = \text{kg/h}$$

$$\text{Efficiency of decorticator} = \frac{Q\text{-}Q_1}{Q_1} \times 100 =$$

$$\text{Breakage of kernels} = \frac{N_2}{N_1 + N_2} \times 100 =$$

Results

The capacity of the decorticator	=	kg/h
The efficiency of decorticator	=	%
Breakage of kernels	=	%

25

Performance Evaluation of The Maize Sheller

Aim

- To determine the efficiency of the maize sheller

Materials required

- Maize sheller
- Maize cobs
- Weighing balance

Theory

Removal of outer sheath is called husking and removal of kernel from maize from maize cob is called shelling. The performance of the sheller is expressed in terms of capacity, percentage of unshelled kernels (corn) and percentage of breakage. The above terms are defined as:

Let,

W_1 = Weight of shelled maize, (kg)

W_2 = Weight of unshelled maize kernels, kg

N_1 = Unbroken kernels in each sample, kg

N_2 = Broken kernels in each sample, kg

T = Time taken for shelling unit weight of maize, h

$$\text{Capacity of the sheller (kg/h)} = \frac{W1}{T}$$

$$\text{Efficiency of the sheller } (\eta) = \frac{W_1 - W_2}{W_1} \times 100$$

$$\text{Breakage of maize kernels (B)} = \frac{N_2}{N_1 + N_2} \times 100$$

Description

The maize sheller consists of a bevel gear fixed adjacent to a shelling disk, which pulls the cob inside a spring loaded tongue, provided above the bevel gear which holes the cob against the shelling disk. The force rotation of shelling disc throws out the empty cobs. The blower separates the inert material from kernels for getting efficient performance. It is advised to sort the cobs manually and adjust the spring loaded nut to the required level depending upon the size of the cob (Fig.25.1). An equipment which remove outer sheath surrounding the maize cob and separate the kernels from desheath cobs in a single operation is known as husker sheller.

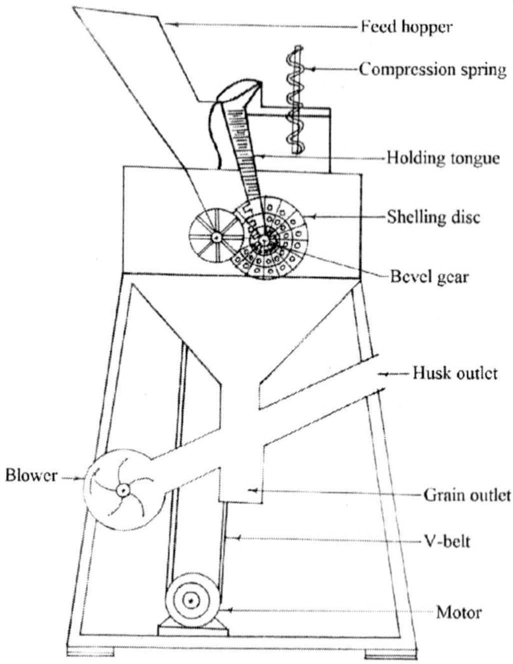

Fig. 25.1. Maize sheller

Procedure

- One kg of maize cob is weighed and fed into the hopper.
- The maize cobs are shelled by operating the shelling unit.
- Time taken for shelling corn is noted.
- The kernel (corn), broken, unshelled cobs are collected at the discharge.
- From the product, kernels (corn), broken and unshelled maize are separated and weighed.

Observations

Time taken for shelling (T) = h

Weight of the shelled maize (W_1) = kg

Weight of unshelled maize (W_2) = kg

Weight of unbroken maize kernels (N_1) = kg

Weight of broken maize kernels (N_2) = kg

Calculation

$$\text{Capacity of the sheller (kg/h)} = \frac{W_1}{T}$$

$$\text{Efficiency of the sheller } (\eta) = \frac{W_1 - W_2}{W_1} \times 100$$

$$\text{Breakage of maize kernels (B)} = \frac{N_2}{N_1 + N_2} \times 100$$

Results

The capacity of the maize sheller = kg/h

The efficiency of the maize sheller = %

Breakage of the maize kernels = %

26

Performance Evaluation of Paddy Parboiling Unit

Aim

- To determine the overall heat utilization for the given paddy parboiling unit.

Materials required

- Lab model parboiling tank
- Soaked paddy
- Thermometer
- Weighing balance

Theory

Paddy parboiling is a hydrothermal process where the paddy is partially boiled to gelatinize the starch. During the gelatinization process the void of the rice kernel are filled and become hard on drying resulting minimum breakage and maximum head rice recovery. The nutrients are also absorbed into the inner mass and loss of nutrient is avoided. In general, the steps are involved in parboiling process are (i) Soaking (ii) Steaming (iii) Drying.

Soaking

The process of water absorption is known as soaking or steeping. Water is penetrates into the center of the grain during soaking. As a result of water absorption the paddy swells. The rate of soaking is dependent on temperature of soaking.

Steaming

Steam is used to complete the gelatinization process as it does not remove moisture from the paddy rather it adds moisture by condensation. If the starch is not fully gelatinized, steaming should be sufficient to cause complete gelatinization.

During steaming the spread of the water-soluble substances inside the paddy grain which is already begun during soaking is continued and increased. The granular texture of endosperm becomes pasty due to gelatinization. Any crack in the endosperm is sealed.

Drying

The parboiled paddy is required to be dried to moisture of 14-16% to obtain the desirable milling and storing properties. Drying of parboiled paddy may be done in the shade, in the sun, or with hot air.

Description

The laboratory modal paddy parboiling tank is made of galvanized iron sheet. The diameter and height of the unit are 30 cm and 40 cm respectively. The distributor assembly of the drum is made of perforated bottom and with perforated main and lateral pipes. Below this distributor assembly about 2-3 kg of water can be stored for steam production. The distributor assembly can hold 3-4 kg of soaked paddy for parboiling. The water reservoir is provided with one KW immersion heater to heat the water and supply steam for parboiling the paddy. The water available in the reservoir is heated by the heater and the steam produced is distributed to the soaked paddy through the distributor assembly (Fig.26.1). The distributed steam will steam the paddy and thus the paddy is parboiled.

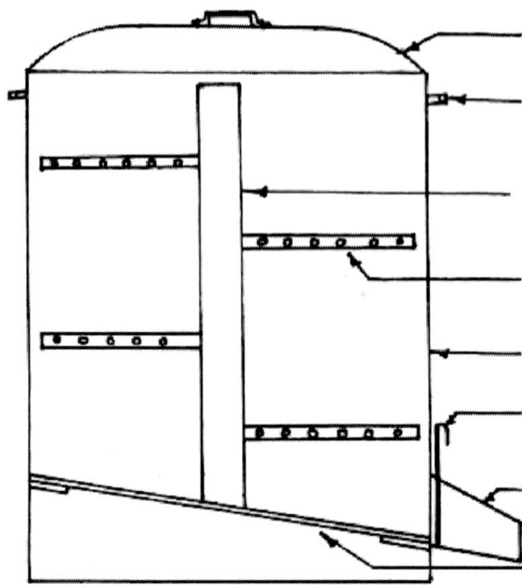

Fig. 26.1. Schematic diagram of lab model paddy parboiling unit

Procedure

- The water reservoir of the tank is filled with known volume of water.
- The initial temperature of water was noted.
- The distributor assembly was placed and filled with soaked paddy after tanking its weight.
- The initial temperature of paddy was also noted.
- The heater is connected to the mains through watt meter and switched on.
- The time of start of experiment is noted and completion of process is checked.
- The splitting of husk is an indication for the completion of the process.
- The time and watt meter readings are noted.
- The final temperature of water and parboiled paddy are recorded.
- Initial moisture content before soaking and moisture after soaking was found to the dry weight of paddy.

Observation and calculation

Initial weight of water taken	(W_1)	=	kg
Initial temperature of water	(T_1)	=	°C
Specific heat of water	(Cp_w)	=	Kcal/kg/°C
Weight of paddy	(W_2)	=	kg
Initial temperature of paddy	(T_2)	=	°C
Specific heat of dry paddy	(Cd_p)	=	Kcal/kg°C
Final temperature of water	(T_3)	=	°C
Final temperature of paddy	(T_4)	=	°C
Final weight of water	(W_3)	=	kg
Weight of wet paddy	(W_4)	=	kg
Time taken for parboiling	(t)	=	h
Power consumed	(P)	=	KW
Heat supplied (H_s)	$P \times 860 \times t$	=	Kcal

Heat utilized (H_u) = (Heat utilized for sensible heat) + (latent heat) + (steaming of the paddy)

$$= [Cp_w \times (W_1 - W_3) \times (T_3 - T_1)] + [(W_1 - W_3)\lambda] + [Cd_p \times W_2 \times (T_4 - T_2)] + [Cp_w \times (W_4 - W_2) \times (T_4 - T_2)]$$

$$\text{Overall heat utilization factor} = \frac{\text{Heat utilized, } H_u}{\text{Heat supplied, } H_s}$$

Result

The overall heat utilization factor for the given paddy-parboiling unit is____________________.

27

Study of Shelling Efficiency of The Rubber Roll Paddy Dehusker

Aim

- To study on working principle and shelling efficiency of rubber-roll paddy dehusker.

Materials required

- Rubber-roll paddy dehusker
- Paddy sample
- Electronic weighing balance

Construction

Fig. 27.1. Lab model rubber-roll dehusker

Rubber-roll Sheller consists of two rubber rolls rotating in opposite direction at different speeds. One of the roll is fixed while the other is adjustable to obtain desired clearance between them. The rolls are driven by mechanically. The adjustable roll normally runs about 25% slower than the fixed one. A feeder feeds paddy uniformly to the machine (Fig.27.1). Paddy is fed in thin layer through the clearance between two rubber rolls, rotating in opposite directions at different speed. Difference in speeds of the rolls develops a shearing force on grain surface results in the splitting, breaking and separating the husk from the grains. One part of the husk is subjected to shearing forces whereas the other part in contact with the slower roll is under compression and is thus subjected to breaking forces. Husking is done by the action of these forces. The clearance between them should be smaller than the mean thickness of paddy and may be adjusted subsequently by judging the shelling efficiency. If the gap between rolls is properly adjusted, it can shell up to 95% of paddy fed to it. At decreased gap excess pressure results more breakage of grain and can also cause coloring of shelled rice. During shelling the faster roll should have a peripheral speed of 10 -13 m/s, while the slower roll 8-10 m/s.

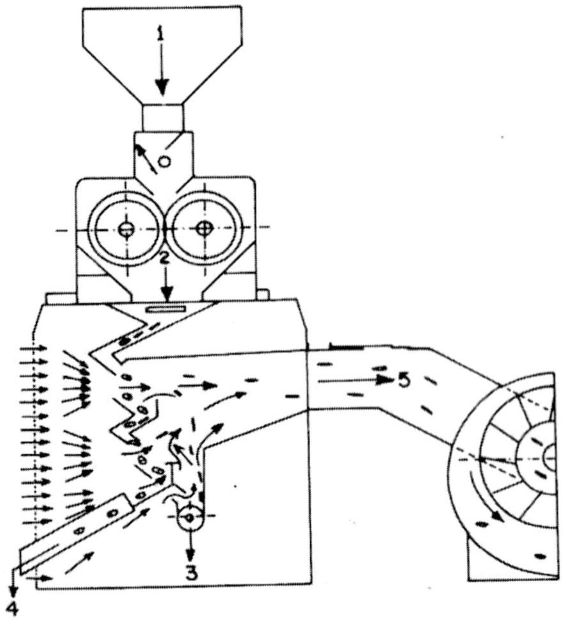

Fig. 27.2. Schematic diagram of rubber-roll sheller with husk separator

1. Paddy, **2.** Mixture of brown rice, paddy and husk, **3.** Immature grains, **4.** Brown rice and paddy, **5.** Husk.

Test Procedure

- Two kilogram of paddy is weighed and transferred to the feed hopper of the rubber roll sheller.
- The motor is switched on and the rollers are allowed to attain the rated speed.
- The feed hopper is opened and the paddy is passed through the roller.
- The stopwatch is used to note the time taken for dehusking the paddy.
- Three samples each weighing 50 grams is taken from feed and the brown rice and husk are separated manually.
- Three samples weighing 50 grams of dehusked sample is taken using the sample divider and unshelled paddy, shelled rice, broken rice and mealy wastes are separated manually.

Observations

n_1 = Weight of paddy taken for shelling/hulling (g)

n_2 = Weight of unshelled/unhulled paddy after hulling (g)

s = Weight of sample taken from the dehusked products (g)

k = Weight of whole head rice after shelling/hulling (g)

d = Weight of broken rice after shelling/hulling (g)

m = Weight of mealy waste in the product after shelling/hulling (g)

g = Weight of husk (g)

T = Time taken for shelling (hr)

Calculations

1. Capacity of the rubber roll sheller

$$C_s = \frac{n_1}{T}$$

2. Head rice recovery

$$H_r = \frac{k-d}{s} \times 100$$

3. Total rice yield

$$R_y = \frac{k}{s} \times n_1$$

4. Percentage whole head rice recovered (%)

$$= \left(1 - \frac{g-n_2}{k} \times 100\right)$$

5. Percentage of brokens (%)

$$= \frac{d}{k} - (g + n_2) \times 100$$

6. Coefficient of hulling: The coefficient of hulling is defined by the amount of unhusked paddy obtained from the total amount of paddy input.

$$E_{hulling} = \left(\frac{n_1 - n_2}{n_1}\right) = \left(1 - \frac{n_2}{n_1}\right)$$

7. Coefficient of wholeness: Coefficient of wholeness of kernel gives the amount of whole head rice out of the total rice obtained. Coefficient of

wholeness of kernel is defined by the proportion of whole kernels to the total amount of kernels (kernel + crushed grain + mealy waste) obtained.

$$Ewk = \frac{k}{k + d + m}$$

8. Effectiveness of hulling/husking: The milling efficiency is defined by the product of coefficient of hulling and coefficient of wholeness of grain.

$$\text{Milling efficiency,\%} = E_{hulling} \times E_{wk} \times 100$$

Results

Shelling efficiency of rubber-roll paddy dehusker is ____________.

28

Design of Bag Storage System

Aim

- To design a bag storage structure for cereals, pulses and oil seeds

Specifications

These structures are generally used for the storing grain in the bags. The length of the structure is about twice the width or greater than that. Special care should be taken to make the side walls, floor and roof in constructing the structure. The walls are made of bricks or stones laid either in lime mortar (1:2) or cement mortar (1:6). The thickness is restricted to a minimum of 37.5 cm or maximum of 45 cm. The height of the walls on which the trusses are placed is kept about 5.5 m. Each of the structures is provided with two large size doors of 2.4 × 2.4 m and top ventilators.

The top ventilator is kept at a height of 60 cm above the floor level. It is also provided with iron rods, wire netting and a shutter. A ground ventilator having an opening of 30 × 30 cm is provided below each corresponding top ventilator. Besides this, a sun shade is provided on both the top and ground ventilators. The roof is either gabled or flat. The gable roof is covered with corrugated metal sheet with maximum precautions taken to make it leak proof. However, a flat roof is more durable as it is made of either reinforced brick or reinforced concrete of about 10 to 12.5 cm thickness.

Stacking arrangement in bag storage

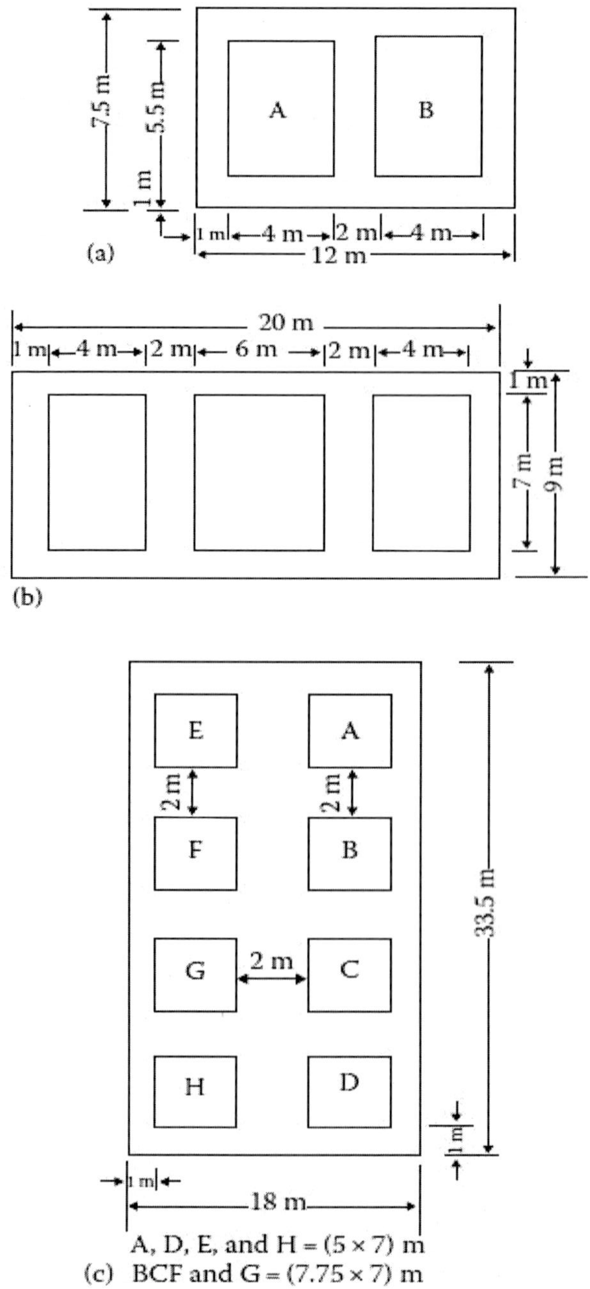

Fig.28.1. Stacking arrangement in bag storage

Bagged food grains are arranged in stacks with a base of 6 m × 10 m with a stack height varying from 3 to 5.5 m, leaving 27 per cent of the floor areas for alley ways. Stacks should be arranged 80 cm away from the wall and leave clear distance between the stacks about 2 m. (Fig.28.1). Generally a godown has a capacity of 5000 tones and consists of 3 to 4 compartments, each having a span of not less than 21.7 m with a clear height of 5.5 m. Air circulations is maintained through steel ventilators and air inlets of rolled steel sections.

Solved problem

Design and draw the plan and elevation in section, of a bag storage structure to store 450 tonnes of paddy. Assume reasonable data where necessary.

Solution

Design capacity of the storage structure = 450 tonnes = 4,50,000 kg

Assume, size of gunny bag = 100 cm long, 60 cm wide, and 30 cm high.

Assume, the capacity of bag of 100 x 60 cm x 30 cm = 75kg

$$\text{Number of bags} = \frac{450000}{75} = 6000$$

Size of stack

Let,

There be 10 bags in length and 10 bags in width in a stack.

Therefore, number of bags/layer = 10 x 10 =100

Let,

There are 12 layers in a stack

Total number of bags/stack = 100 x 12 = 1200

$$\text{Number of stack required} = \frac{6000}{1200} = 12$$

Space required by each stack

Length = 10 x 1.0 = 10 m

Width = 10 x 0.6 = 6 m

Height = 12 x 0.3 = 3.6 m

The clear distance between the walls and the end of stack = 0.8 m

The clear distance between the stacks = 2 m

Length of floor = (5 x 6) + (4 x 2) + (2 x 0.8) = 39.6 m

Width of floor = (1 x 10) + (2 x 0.8) = 11.6 m

Height = 12 x 0.3 = 3.6 m

Leave clear height of 1.4 m above the stack

Total height of wall =3.6 + 1.4 = 5m

The overall dimensions of the godown may be taken as

Length of floor = 40.0 m

Width of floor = 12.0 m

Height of wall = 5.0 m

29

Design of Shallow and Deep Bins

Aim

- To design shallow and deep bins

Theory

- The structure behaves as a deep or shallow bin depending upon its relative dimensions such as depth and diameter or width. It is decided from the concept of the plane of rupture.
- **Plane of rupture**: The plane of rupture is that surface down which a wedge of material bounded by one wall face, the free surface, and the plane of rupture would start sliding if bounding wall were to move.
- **Shallow bin**: A bin whose relative dimensions are such that the plane of rupture meets the grain surface before it strikes the opposite side as shown in Fig. is called shallow bin (Fig.29.1a).
- **Deep bin**: A bin in which plane of rupture meets the opposite side before it emerges from the grain as shown in Fig. is called deep bin (Fig.29.1b).

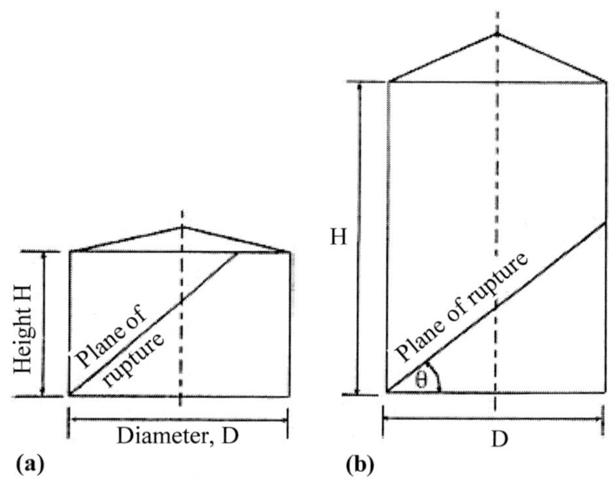

Fig.29.1. Relation between **(a)** shallow bin **(b)** deep bin

1. Rankine formula for shallow bin

- Rankine developed a model to estimate the lateral grain pressure in silos considering shallow bin factor The formula used for estimating grain pressure in shallow bins is:

$$P_l = K\omega h$$

Where,

P_l = lateral pressure exerted by the grain on the bin wall

ω = bulk density of grain

h = depth of grain from the top of the bin surface

f = angle of internal friction of the grain

K = Rankine's earth pressure coefficient

$$K = \frac{1 - \sin\phi}{1 + \sin\phi}$$

If ϕ is the angle of repose of the material then the angle of rupture with respect to horizontal may be taken as follows:

$$\frac{90 + \phi}{2}$$

If H and D be the depth and width of the structure, then the structure can be taken to be shallow bin if

$$H < D \tan\left(\frac{90 + \phi}{2}\right)$$

The structure is deep if

$$H > D \tan\left(\frac{90 + \phi}{2}\right)$$

2. Airy formula for shallow bin

- Airy developed following model for lateral pressure exerted by grains in a shallow bin:

$$P_l = \omega h \left[\frac{1}{\sqrt{\mu(\mu + \mu')} + \sqrt{1 + \mu^2}}\right]^2$$

Where,

P_l = lateral pressure exerted by the grain on the bin wall

ω= bulk density of grain

h = depth of grain to point under consideration

μ = coefficient of friction of grain on the wall

$\mu = \tan \phi$

$\mu' = \tan \phi'$

ϕ = angle of internal friction of grain

ϕ' = angle of internal friction between wall and grain

3. Janssen formula for deep bin

- Janssen developed a model for lateral grain pressure in deep bins by taking into account the friction between the grain and bin wall (Fig.29.2).

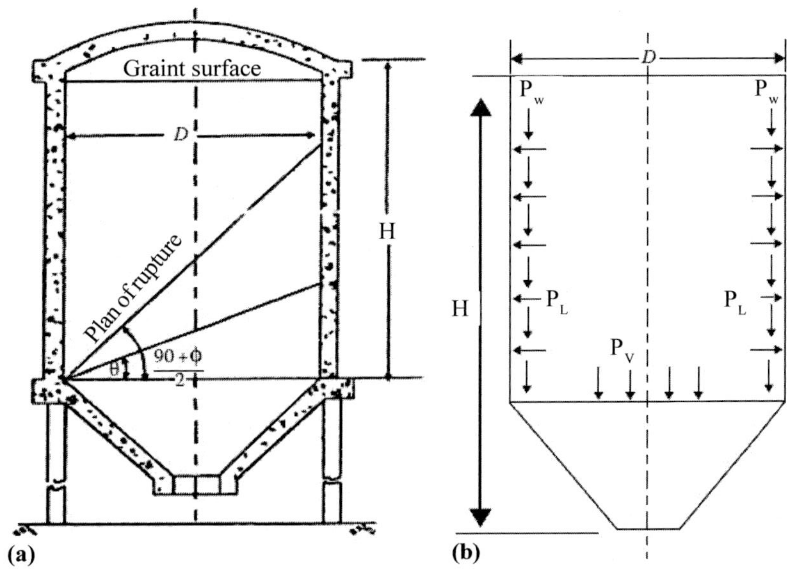

Fig.29.2. Schematic diagram of **(a)** deep bin **(b)** stress analysis in deep bin

The lateral pressure exerted by the grain on the bin wall can be obtained by:

$$P_l = \frac{\omega R}{\mu}\left[1 - e^{-\frac{K\mu h}{R}}\right]$$

Vertical pressure transferred owing to wall friction:

$$P_w = \omega R\left[\frac{hR}{\mu' K}\left(1 - e^{-\frac{K\mu' h}{R}}\right)\right]$$

Where,

P_l = lateral pressure exerted by the grain on the bin wall

P_w = vertical pressure transferred owing to wall friction

P_v = vertical pressure

ω = bulk density of grain

R = hydraulic mean radius (D/4)

μ = coefficient of friction of grain on the wall, $\mu = \tan\phi$

h = depth of grain

K = ratio of lateral pressure to vertical pressure in the grain

ϕ = angle of internal friction of the grain

$$K = \frac{1 - \sin\phi}{1 + \sin\phi}$$

4. Reimbert and Reimbert theory

- As per this theory, ratio of vertical pressure to lateral pressure is varies with depth and shape of the silos.

Let,

H = total height of the bin

Z = height of upper cone

h = height of grain against the wall = H - Z

r = radius of cylindrical bin, D/2

ϕ = minimumangle of internal friction of the grain

R = hydraulic mean radius = D/4

μ = minimum angle of friction on smooth sheeting

K = ratio of horizontal and vertical pressure intensity

A = characteristic abscissa in the silo

P_{max} = maximum lateral pressure of a bin

L = total load on the bottom of the silo

S = cross-sectional area of the silo

ω = bulk density of the grain

The ratio of horizontal and vertical pressure intensity:

$$K = \frac{1-\sin\phi}{1+\sin\phi} = \tan^2\left(\frac{\pi}{4} - \frac{\phi}{2}\right)$$

The height of upper cone of stored material is given by; $Z = r \tan\phi$

The value of characteristic abscissa in the bin:

$$A = \frac{R}{\mu K}$$

Maximum lateral pressure of a bin:

$$P_{max} = \frac{\omega_{max} \times R}{\mu}$$

Total load at the bottom of the silo:

$$L = S\omega\left[h\left(\frac{h}{A}+1\right)^{-1} + \frac{Z}{3}\right]$$

Problem 1: Paddy is loaded in a concrete silo having an internal diameter 5 m and clear height 7 m. The bulk density of the paddy is 700 kg/m^3, the angle of internal friction for paddy is 30° and that for concrete and wheat is 25°. Estimate the lateral pressure exerted by the paddy using Rankine theory.

P_l = lateral pressure exerted by the grain on the bin wall ?

ω = bulk density of paddy = 700 kg/m^3

h = depth of grain = 7 m

D = internal diameter = 5 m

ϕ= angle of internal friction of the grain = 30°

K = Rankine's earth pressure coefficient

The following relationship is used to check for the bin is shallow

$$H < D \tan\left(\frac{90+\phi}{2}\right)$$

$$H < 5 \tan\left(\frac{90+30}{2}\right)$$

H < 8.66, hence it is a shallow bin

Rankine formula for lateral pressure of the grain: $P_l = K\omega h$

$$P_l = \left(\frac{1-\sin\phi}{1+\sin\phi}\right) \times \omega \times h$$

$$P_l = \left(\frac{1-\sin 30}{1+\sin 30}\right) \times 700 \times 7 = 1633.33 kg/m^2$$

Problem 2: Calculate the maximum lateral pressure exerted by wheat grain at the bottom of a circular concrete silo using Airy theory. Internal diameter of the silo is 3 m and clear height is 8 m. The bulk density of the wheat grain is 900 kg/m³, the angle of internal friction for wheat is 25° and that for concrete and wheat is 24°.

P_l = lateral pressure exerted by the grain on the bin wall ?

ω = bulk density of the wheat grain = 900 kg/m3

h = height of the bin = 8 m

d = internal diameter of the bin = 3 m

ϕ = angle of internal friction for wheat = 25°

ϕ' = angle of friction between grain and wall = 24°

μ = coefficient of friction of grain on the wall

μ = tan ϕ = tan (25°) = 0.466

μ' = tan ϕ' = tan (24°) = 0.445

Airy formula estimating lateral pressure against a shallow bin wall is:

$$P_l = \omega h \left[\frac{1}{\mu(\mu+\mu') + \sqrt{1+\mu^2}}\right]^2$$

$$P_l = 900 \times 8 \left[\frac{1}{\sqrt{0.466(0.466+0.455)} + \sqrt{1+0.466^2}}\right]^2$$

$$P_l = 900 \times 8 \left[\frac{1}{0.6515+1.1031}\right]^2 = 2338.5 kg/m^2$$

Problem 3: A RCC cylindrical grain storage bin has internal diameter of 5 m and is 20 m deep. It is completely filled with paddy weighing 600 kg/m³. The angle of internal friction for paddy can be taken as 35 degree while the angle of friction between paddy and bin wall is 30 degree. The ratio of horizontal and vertical pressure intensity is 0.4. Calculate the lateral pressure intensity at 2 m interval using Janssen formula.

P_l = lateral pressure exerted by the grain on the bin wall ?

ω = bulk density of the paddy = 600 kg/m^3

h = height of the bin = 20 m

d = internal diameter of the bin = 5 m

ϕ = angle of internal friction for paddy = 35°

ϕ' = angle of friction between grain and wall = 30°

μ = coefficient of friction of grain on the wall

$\mu = \tan\phi = \tan(30°) = 0.577$

R = hydraulic mean radius = D/4 = 5/4 = 1.25 m

K = ratio of horizontal and vertical pressure intensity = 0.4

Janssen formula for lateral pressure exerted by the grain on the deep bin wall:

$$P_w = \frac{\omega R}{\mu}\left[1 - e^{-\frac{K\mu h}{R}}\right]$$

Lateral pressure (P_l) at 2 m grain depth,

$$P_l = \frac{600 \times 1.25}{0.577}\left[1 - e^{-\frac{0.4 \times 0.577 \times 2}{1.25}}\right] = 401.6 \text{kg}/\text{m}^2$$

Lateral pressure (P_l) at 4 m grain depth,

$$P_l = \frac{600 \times 1.25}{0.577}\left[1 - e^{-\frac{0.4 \times 0.577 \times 4}{1.25}}\right] = 679.8 \text{kg}/\text{m}^2$$

Lateral pressure (P_l) at 6 m grain depth,

$$P_l = \frac{600 \times 1.25}{0.577}\left[1 - e^{-\frac{0.4 \times 0.577 \times 6}{1.25}}\right] = 870.8 \text{kg}/\text{m}^2$$

Lateral pressure (P_l) at 8 m grain depth,

$$P_l = \frac{600 \times 1.25}{0.577}\left[1 - e^{-\frac{0.4 \times 0.577 \times 8}{1.25}}\right] = 1003 \text{kg}/\text{m}^2$$

Lateral pressure (P_l) at 10 m grain depth,

$$P_l = \frac{600 \times 1.25}{0.577}\left[1 - e^{-\frac{0.4 \times 0.577 \times 10}{1.25}}\right] = 1094.4 \text{kg}/\text{m}^2$$

Lateral pressure (P_l) at 12 m grain depth,

$$P_l = \frac{600 \times 1.25}{0.577}\left[1 - e^{-\frac{0.4 \times 0.577 \times 12}{1.25}}\right] = 1158 \text{kg}/\text{m}^2$$

Lateral pressure (P_l) at 14 m grain depth,

$$P_l = \frac{600 \times 1.25}{0.577}\left[1 - e^{-\frac{0.4 \times 0.577 \times 14}{1.25}}\right] = 1202\,kg/m^2$$

Lateral pressure (P_l) at 16 m grain depth,

$$P_l = \frac{600 \times 1.25}{0.577}\left[1 - e^{-\frac{0.4 \times 0.577 \times 16}{1.25}}\right] = 1232.2\,kg/m^2$$

Lateral pressure (P_l) at 18 m grain depth,

$$P_l = \frac{600 \times 1.25}{0.577}\left[1 - e^{-\frac{0.4 \times 0.577 \times 18}{1.25}}\right] = 1253\,kg/m^2$$

Lateral pressure (P_l) at 20 m grain depth,

$$P_l = \frac{600 \times 1.25}{0.577}\left[1 - e^{-\frac{0.4 \times 0.577 \times 20}{1.25}}\right] = 1268.6\,kg/m^2$$

Problem 4: Using Reimbert and Reimbert theory, calculate the load on the bottom, and the lateral thrust at every 2 m depth on the cylindrical silo of 2.5 m diameter and 20 m in height. The silo is made of steel with smooth walls and it is filled with wheat. The characteristics of stored wheat are as follows: minimum bulk density = 720 kg/m^3, maximum bulk density = 830 kg/m^3, minimum angle of internal friction = 25°, maximum angle of internal friction = 30°, minimum angle of friction on smooth sheeting = 18°, angle of repose of the wheat = 25°.

r = radius of cylindrical bin = D/2 = 2.5/2 = 1.25 m

ϕ = minimum angle of internal friction = 25°

H = total height of the bin = 20 m

R = hydraulic mean radius = D/4

μ = minimum angle of friction on smooth sheeting

K = ratio of horizontal and vertical pressure intensity

A = characteristic abscissa in the silo

P_{max} = maximum lateral pressure of a bin

L = total load on the bottom of the silo

h = height of grain against the wall

S = cross-sectional area of the silo

ω = bulk density of the grain

R = D/4 = 2.5/4 = 0.625 m

$\mu = \tan \phi = \tan (18^o) = 0.3249$

$h = H - Z = 20 - 0.5825 = 19.417$ m

Height of the upper cone of stored material is given by:

$Z = r \tan \phi = 1.25 \tan 25^o = 1.25 \times 0.4663 = 0.5825$ m

Ratio of horizontal and vertical pressure intensity:

$$K = \frac{1 - \sin\phi}{1 + \sin\phi} = \tan^2\left(\frac{\pi}{4} - \frac{\phi}{2}\right) = \tan^2\left(\frac{\pi}{4} - \frac{25}{2}\right) = 0.4058$$

Characteristic abscissa in the silo:

$$A = \frac{R}{\mu K} = \frac{0.625}{0.3249 \times 0.4058} = 4.74\text{m}$$

Maximum lateral pressure of a bin:

$$P_{max} = \frac{\omega_{max} \times R}{\mu} = \frac{830 \times 0.625}{0.3249} = 1596.6\,\text{kg}/\text{m}^2$$

Total load at the bottom of the silo:

$$L = S\omega\left[h\left(\frac{h}{A} + 1\right)^{-1} + \frac{Z}{3}\right]$$

$$L = \frac{\pi \times 2.5^2}{4} \times 830\left[19.417\left(\frac{19.417}{4.74} + 1\right)^{-1} + \frac{0.5825}{3}\right] = 16314.8\text{kg}$$

30

Design of Screw Conveying System

Aim

- To design a screw conveyor

Theory

- The screw conveyor consists of screw blade, screw shaft, coupling, tubular trough, cover, inlet and outlet gates, bearings and drive mechanism.
- A shaft with spiral screw revolves in the tubular trough. The screw shaft is supported by end and hanger bearings. The rotation of screw pushes the grain along the trough.
- The screw blade or flight is a continuous one-piece helix, shaped from a flat strip of steel welded on to the shaft.
- The screw shaft is usually a joint less tube with thick sides and a high tensile strength to reduce the weight.
- The tubular trough is completely closed with circular cross-section and mostly used for conveying materials at inclination or for vertical lift.
- The screw conveyor is designed for clockwise or counter clockwise rotation. The change in direction of rotation does not affect the capacity.
- The capacity of screw conveyor is influenced by the screw diameter, inclination of the screw blade, speed of the blade, shaft diameter and cross-section of loading.
- The pitch of the standard screw which is the distance from the centre of one thread to the centre of the next thread is equal to its diameter.

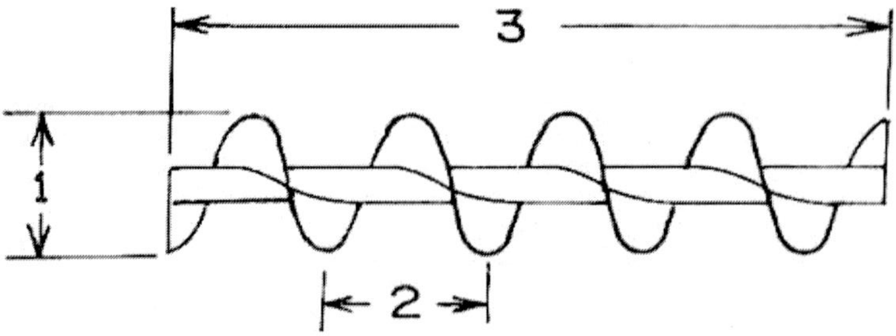

Fig 30.1. Screw conveyor details

1. Screw diameter **2.** Pitch of screw **3.** Screw length

Theoretical conveyance capacity

The theoretical conveyance capacity of the screw conveyor can be given by the following equation.

$$Q_{tc} = \frac{\pi}{4}(D^2 - d^2) \times P \times N$$

Where,

Q_{tc} = Theoretical capacity of screw conveyor, m^3/min

D = Screw diameter, m

d = shaft diameter, m

P = Pitch length, m

N = Speed of screw conveyor, rpm

Actual capacity

The theoretical capacity is more than the actual capacity because of the screw housing clearance and loading factor (Fig. 30.2).

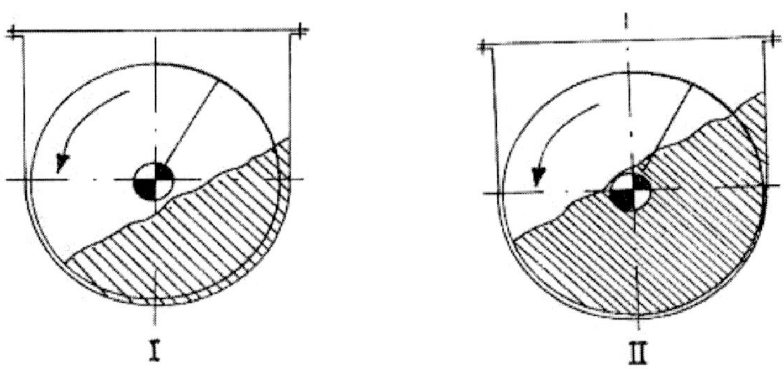

I. 30% loading **II.** 55% loading

Fig 30.2. Screw conveyor loading factor

Actual capacity can be determined by product of theoretical capacity and bulk material weight.

$Q_{ac} = Q_{tc} \times W$ (kg/min)

Where,

Q_{ac} = Actual capacity of the screw conveyor, kg/min

Q_{tc} = Theoretical capacity of the screw conveyor, m^3/min

W = Bulk weight of the materials, kg/m^3

Power requirement

The power requirement of screw conveyors for horizontal operation may be determined by:

$$HP = \frac{Q \times L \times K}{4500}$$

Where,

Q = Actual capacity of conveyor, kg/min

L = Conveyor length, m

K = Material factor, (for paddy 0.4)

Problem

A screw conveyor mounted on a 4 cm diameter shaft. The diameter of the screw pitch and screw diameter both equal to 30 cm. Estimate its actual capacity of conveying wheat weighing 850 kg/m3 while operating at 150 rpm. Assume actual capacity as 50 per cent of theoretical capacity. What horsepower motor will be required for a screw length of 8 m and the material factor for wheat is 0.4?

Solution

Given,

D = screw diameter = 30 cm

d = shaft diameter = 4 cm

P = screw pitch length = 30 cm

N = speed of screw conveyor = 150 rpm

L = length of screw = 8 m

K = material factor = 0.4

ω = bulk density of wheat = 850 kg/m^3

Theoretical capacity (Q_t)

$$Q_{tc} = \frac{\pi}{4}(D^2 - d^2) \times P \times N \, (m^3 / min)$$

$$Q_{tc} = \frac{3.14}{4} \times (0.3 - 0.04^2) \times 0.3 \times 150 = 3.123 \, m^3 / min$$

Actual capacity (Q_{ac})= theoretical capacity (Q_{tc}) × bulk density of the material (w)

$$Q_{tc} = \left(3.123 \times \frac{50}{100} \right) \times 850 = 1327.16 \, kg/min$$

According to Henderson, horsepower of the screw conveyor

$$HP = \frac{Q \times L \times K}{4500}$$

$$HP = \frac{1327.16 \times 8 \times 0.4}{4500} = 0.944$$

31

Design of Bucket Elevator

Aim

- To design a bucket elevator

Theory

- The bucket elevator's capacity mainly depends on bucket size, conveying speed, bucket design and spacing, the way of loading and unloading, the bucket and the characteristic of bulk material.
- Bucket elevators with a belt carrier can be used at fairly speeds of 2.5 to 4 m/s. The speed of the belt depends on the head pulley.
- If the belt speed is too low, the discharge of the grains become more difficult, with too high speed the buckets are not fed well.
- The theoretical horsepower should be increased 10-15% to provide for friction, power requirements for loading, power transmission and drive losses.
- Actual horsepower requirement is 10-15% of the theoretical horsepower.
- When a product mass turns around a head pulley, it is influenced by two forces,
- (i) Gravitational force, which is oriented downwards and
- (ii) Centrifugal force, which is oriented outwards.

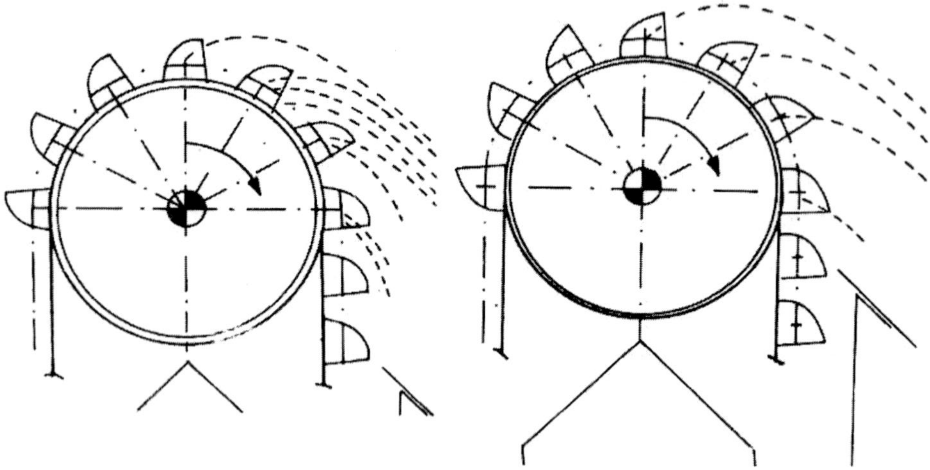

Fig.31.1. Bucket elevator discharge methods
1. Low speed gravitational discharge 2. High speed centrifugal discharge

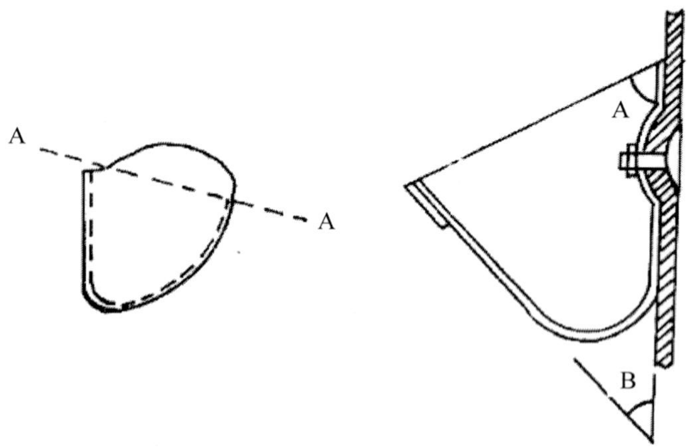

Fig.31.2. Bucket configuration

The magnitude of the centrifugal force that is oriented outward (Fig.31.1)

Centrifugal force

$$C_f = \frac{W \times V^2}{g \times r}$$

Where,

C_f = Centrifugal force

W = Weight of grain, kg

V= Velocity of product mass, m/s

g = Acceleration due to gravity, m^2/sec

r = Radius of the head wheel plus one-half of the projection of the bucket, m

Speed of the head pulley

For calculation of the speed of head pulley and optimum centrifugal discharge, the resultant force is zero. It means that the centrifugal force (C_f) is equal to the force of gravity (W).

Hence, $C_f = W$

$$W = \frac{W \times V^2}{g \times r}$$

$$V = \sqrt{g.r}$$

$$\text{Since Velocity}, V = \frac{2\,\pi\,n\,r}{60}$$

n = speed of the drive pulley, rpm

$$\sqrt{gr} = \frac{2\pi nr}{60}$$

Simplify, we get

$$n = \frac{60\sqrt{g}\,\sqrt{r}}{2\pi\sqrt{r}\,\sqrt{r}} = \frac{29.9}{\sqrt{r}}$$

Capacity of the bucket elevator

The bucket elevator's capacity may be calculated by the following equation:

$$Q = C \times N \times S \times \rho$$

Where,

Q = Elevator capacity, kg/min

C = Bucket's capacity, m^3

N = Number of buckets lifted/sec

S = Speed of the belt, m/sec

ρ = Material density, kg/m^3

Horsepower requirement

The theoretical horsepower requirement for the bucket elevator can be calculated by the following equation:

$$HP = \frac{QHF}{4500}$$

Where,

Q = capacity of the bucket elevator, kg/min

H = height or lift of the elevator, m

F = material factor

(1.5 for elevators loaded on the up side and 1.2 for elevators loaded on the bottom side)

Problem

In a bucket elevator for lifting paddy, each bucket is 25 cm long and has a cross-section which is a section of a circle having a radius of 15 cm and substending an angle of 81 at the centre. The buckets are spaced 45 cm apart, the lift is 25 m and the head wheel has a diameter of 60 cm. Calculate

a) Belt speed so that the discharge is centrifugal

b) Capacity of lifting paddy which weight 580 kg/m^3 and

c) Horsepower required assuming an overall efficiency of 85 per cent.

Given,

b = length of the bucket = 25 cm

R = radius of curvature of bucket = 15 cm

θ = angle substended by the segment in radian = 81° = 0.7065 radians

d = diameter of the head wheel = 60 cm (or) radius, r = 30 cm

h = height of lift = 25 m

x= spacing between the buckets = 45 cm

ρ = bulk weight of the paddy = 580 kg/m^3

Elevator capacity

$$Q = C \times N_b \times S \times \rho$$

Where,

Q = Elevator capacity, kg/min

C = Bucket's capacity, m^3

N = Number of buckets lifted/sec

S = Speed of the belt, m/sec

ρ = Material density, kg/m^3

Belt speed

In order to have centrifugal discharge, the centrifugal force (C_f) and weight of grain (W) must be equal at a point near the top of the travel

$$V = \sqrt{g.r}$$

$$V = \sqrt{9.81 \times 0.30} = 1.7155\ m/sec$$

Capacity of one bucket (C)

Volume of one bucket = area of bucket × length of bucket

$= (R^2\ \theta) \times b$

$= (0.15)^2 \times 0.7065 \times 0.25$

$= 3.98 \times 10^{-3}\ m^3$

Number of buckets lifted/sec (N)

$$\text{No. of buckets lifted/sec} = \frac{\text{belt speed}}{\text{spacing of buckets}} = \frac{1.715}{0.45} = 4$$

Volume of grain lifted/sec = $C \times N$

$= (3.98 \times 10^{-3}) \times 4$

$= 15.92 \times 10^{-3}\ m^3/sec$

Elevator capacity (Q)

Elevator capacity = Volume of grain lifted × bulk density of the grain

$= (15.92 \times 10^{-3}) \times 580 \times 60$

$= 554.016$ kg/min

Horsepower required (HP)

$$HP = \frac{Q \times H \times F}{4500}$$

Where,

Q = capacity of the bucket elevator, kg/min

H = height or lift of the elevator, m

F = material factor (F = 1.2, for elevators loaded on the bottom side)

$$HP = \frac{554 \times 25 \times 1.2}{4500} = 3.7$$

Horsepower input at an efficiency of 85% = 3.7/0.85 = **4.35 HP**

32

Design of Belt Conveying System

Aim

- To design a belt conveying system

Theory

- The design of belt conveying system is based on available space, horizontal conveying length, conveying lift, characteristics of the materials to be conveyed and capacity requirement.
- On the basis of overall requirement and information, the belt width, belt speed, required horsepower, maximum belt tension and breaking strength of the belt, diameter of the pulleys and idlers and quantity of belt (thickness) are determined to design a belt conveyor (Fig.32.1).
- The first step in the design of a belt conveyor with a specified conveying capacity is to determine the speed and width of the belt. The belt speed should be selected to minimize product spillage. For transportation of the grains, the belt speed should not exceed 3.5 m/s.
- Generally, for grain conveying, belt speed of 2.5 to 2.8 m/s is recommended. The selection of belt width will depend upon the capacity requirement, speed of operation, angle of inclination of belt conveyor, trough angle and depth.
- A trough angle of 20 is best suited for paddy and most other grains. Other common trough angles are 30 and 45 Paddy forms a surcharge angle of 20. Other common surcharge angle ranges between 5 and 30 (Fig.32.2).

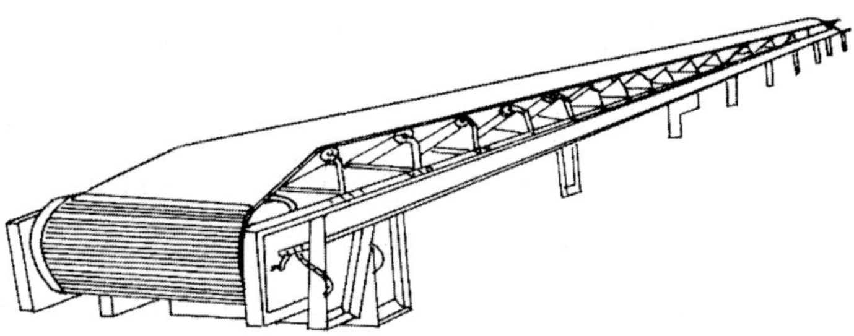

Fig.32.1. Belt conveyor

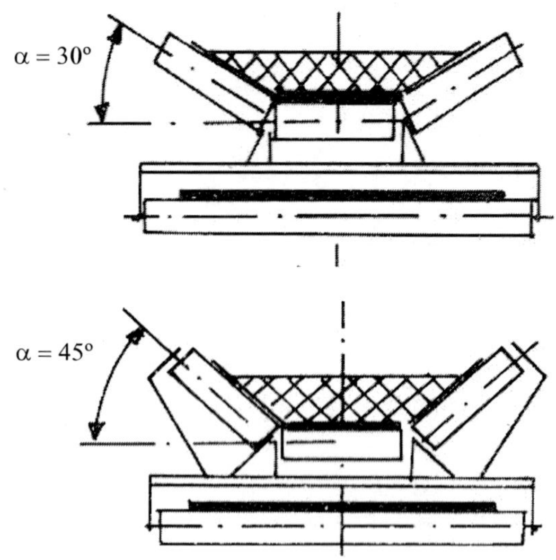

Fig. 32.2. Troughing configurations

Capacity of belt conveyor

$C = A \times S \times \rho$ (kg/min)

Where,

A = cross-sectional area of the belt, m^2

S = speed of the belt, m/min

ρ = bulk density of the grain, kg/m^3

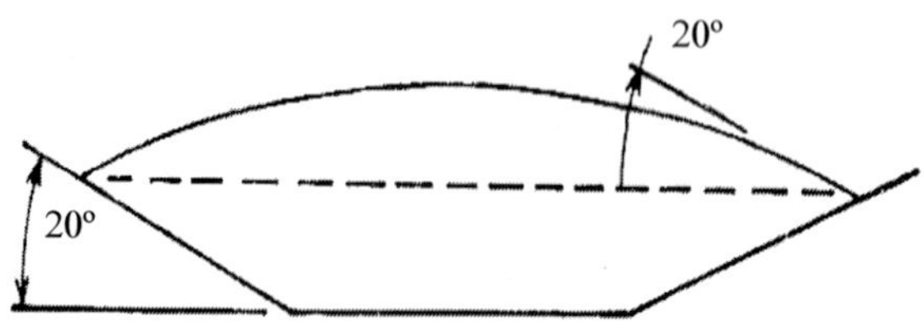

Fig. 32.3. Area of the belt

Power requirement

The horsepower required for operation of belt conveyor for conveying grains can be calculated by the following equation:

$$HP = C \times \left(\frac{A \times LB}{100}\right)$$

Where,

C = Capacity of the belt conveyor, tone/h

L = Length of the belt, m

A & B = constants, depends on the belt width

Table. Values of constants A and B

Belt width, cm	Constants		Additional horsepower for tripper
	A	B	
36	0.2	0.0014	0.7
41	0.25	0.0014	0.85
46	0.3	0.0016	1
50	0.3	0.0018	1.4
60	0.36	0.0022	1.7
76	0.48	0.0029	2.5

Problem

A troughed belt 90 cm wide conforms to a set of three pulleys such that the belt incline is 20° to enable transport of paddy weighing 600 kg/m^3. The belt should have a clear margin of 5 cm on both side and the surcharge angle for paddy is measured to be 20.

i) Calculate the carrying capacity of the belt while moving at a speed of 200 m per min

ii) What would be the horsepower if constants A and B are 0.48 and 0.01 respectively?

Solution

Fig. 32.4 presents the cross section of loaded belt showing the surcharge material.

Cross-section of loaded belt = area of the figure ABCED = area of trapezium ABCD + area of arc DCE

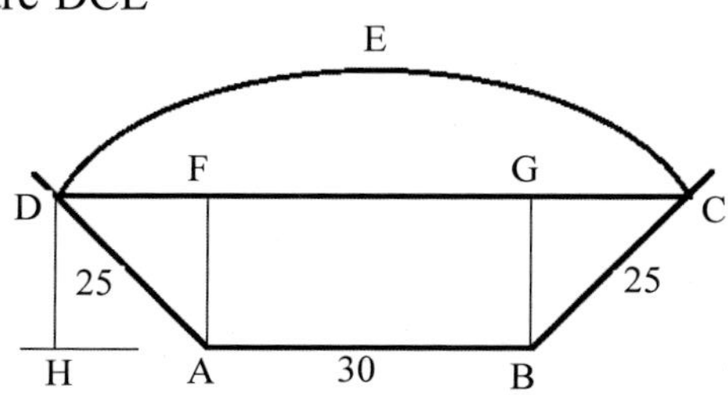

Fig. 32.4. Cross section of loaded belt

Area of the trapezium ABCD

Bottom of the trapezium = 30 cm

Leaving a clearance = 5 cm

AD = BC = (30 – 5) = 25cm

Height of the trapezium = AF = DH

= AD sin 20° = 25 × 0.31 = 7.725 cm

Length of DF = AH = GC

= AD cos 20°

= 25 × 0.951 = 23.776 cm

Top width (DC) = FG + DF + GC

= 30 + 23.776 + 23.776 = 77.55 cm

Area of the trapezium ABCD = (b + t) d/2

= (30 + 77.55) × 7.725/2 = 415.4 cm^2

Area of arc DCE

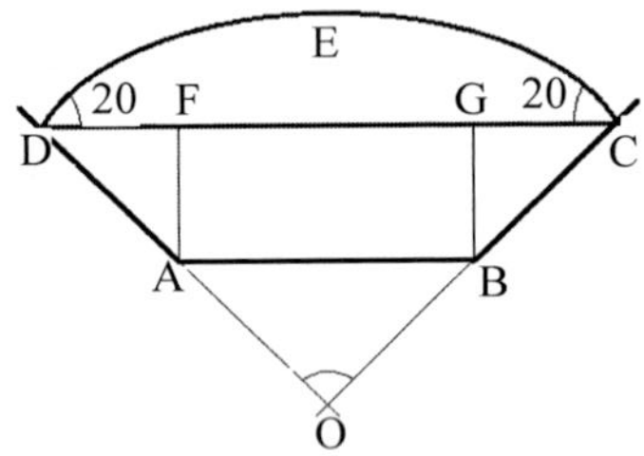

Surcharge angle for paddy is 20°

ie, angle DCE and angle CDE = 20°

Draw lines DO and CO to meet at centre point 'O'

ODEC is a part of the circle having a radius of OD or OC

Area of arc DCE = Area of sector ODEC – area of triangle ODC

Hence area of ODEC = OC^2 × angle DOC in radians

From the geometry,

Angle ODC and angle OCD = 70°

Angle DOC = 40°

Since, DC = 77.55 cm and OCD = 70°

$$\cos\theta = \frac{DC}{2} \times \frac{1}{OC}$$

$$OC = \frac{77.55}{2 \times \cos 70^\circ} = 113.37\,\text{cm}$$

Area of ODEC = $OC^2 \times 40 \times \pi/360$

Thus area of ODEC = $(113.37)^2 \times 0.349 = 4485.61\ cm^2$

$$\text{Area of triangle ODC} = \frac{DC}{2} \times OC \times \sin\theta$$

$$\text{Area of } \Delta \text{ ODC} = \frac{77.55}{2} \times 113.37 \times \sin 70^\circ = 4130.82\ cm^2$$

Area of arc DCE = Area of sector ODEC – area of triangle ODC

= $4485.61 - 4130.82 = 354.79\ cm^2$

Total area of ABCDE = $415.4 + 354.79 = 770.19\ cm^2$

Cross-sectional area of the belt (A) = $0.077019\ m^2$

Speed of the belt (S) = 200 m/min

Bulk density of the grain (ρ) = $600\ kg/m^3$

The capacity of belt conveyor (C) = $A \times S \times \rho$

= $0.077019 \times 200 \times 600$

= 9242.28 kg/min

$$\text{Capacity of belt conveyor, tones/h} = \frac{9242.28 \times 60}{1000} = 554.54 \text{ tones/h}$$

Horsepower required for operation of belt conveyor

Capacity of the belt conveyor (C) = 554.54 tones/h

Length of the belt (L) = 200 m

$A = 0.48$ & $B = 0.01$

$$HP = C \times \frac{A + LB}{100}$$

$$HP = 554.54 \times \frac{0.48 + (200 \times 0.01)}{100} = 13.75 \approx 14\,HP$$

33

Determination of Conveying Efficiency of Screw Conveyor

Aim

- To determine the conveying efficiency of screw conveyor.

Required apparatus

- Screw conveyor
- Tachometer
- Weighing balance
- Stopwatch
- Grains

Description

Screw conveyor is one of the equipment used in material handling to convey the material horizontally. The screw conveyor consists of screw blade, screw shaft, coupling, tubular trough, cover, inlet and outlet gates, bearings and drive mechanism. A shaft with spiral screw revolves in the tubular trough. The screw shaft is supported by end and hanger bearings. The rotation of screw pushes the grain along the trough. The pitch of a stranded screw which is the distance from the center of one thread to the center of the next thread is equal to its diameter (Fig.33.1).

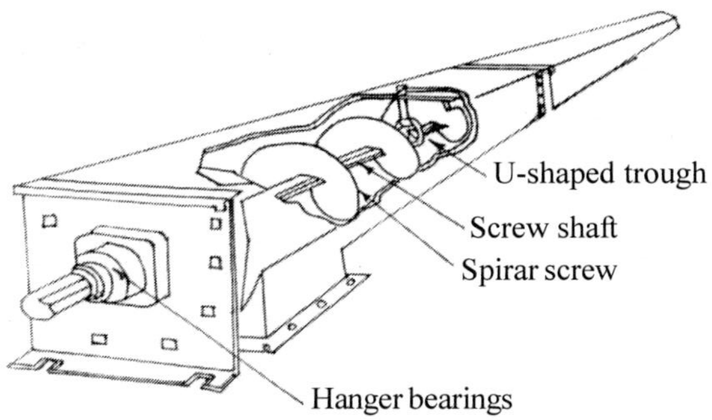

Fig. 33.1. Screw conveyor

For example a 10 cm diameter screw has a pitch of 10 cm. The screw blade or flight is a continuous one-piece helix, shaped from a flat strip of steel welded on to the shaft. The screw shaft is usually a joint less tube with thick sides and a high tensile strength to reduce the weight. The tubular trough is completely closed with circular cross-section and mostly used for conveying materials at inclination or for vertical lift.

The screw conveyor is designed for clockwise or counter clockwise rotation. The direction of rotation does not affect the capacity. The capacity of screw conveyor is influenced by the diameter, inclination of the screw blade, speed of the blade, shaft diameter and cross-section of the loading. For operational reason, gap is provided between the edge of the screw blade and the trough walls. Due to this gap, it is not possible to completely empty the trough of a horizontal screw conveyor.

Test procedure

- The grain is loaded in the feed hopper which is kept at the feeding section of the equipment and the unit is switched on.
- The rpm of the head pulley is measured using a tachometer.
- The grains are fed into conveyor and time is noted.
- The conveyor is run for a particular time and the material collected at the outlet is weighed.

Observations

Bulk density of the material	(ω)	=	kg/m^3
Screw diameter	(D)	=	m
Shaft diameter	(d)	=	m
Pitch of the screw	(P)	=	m
Speed of the rotation	(N)	=	rpm
Length of the conveyor	(L)	=	m
Weight of grain collected	(W_g)	=	kg
Time taken	(T)	=	h
Material factor	(K)	=	0.4 for paddym

Calculations

$$\text{Theoritical capacity}, Q_{tc} = \frac{\pi}{4}(D^2 - d^2) \times P \times N \, (m^3/min)$$

$$\text{Actual capacity } (Q_{ac}) = \frac{W_g}{T} = \text{kg / min}$$

(OR)

$$\text{Actual capacity } (Q_{ac}) = Q_{tc} \times \omega \text{ (kg/min)}$$

$$\text{Conveyance efficiency} = \frac{Q_{ac}}{Q_{tc}}$$

$$\text{Horse power} = \frac{Q_{ac} \times L \times K}{4500}$$

Result

i) The conveyance efficiency of the screw conveyor is ___________.

ii) The power requirement of screw conveyors for horizontal operation is_________________.

34

Determination of Conveying Efficiency of Bucket Elevator

Aim

- To determine the efficiency of bucket elevator.

Materials required

- Lab bucket elevator
- Tachometer
- Weighing balance
- Stopwatch
- Grains

Description

Bucket elevator is one of the equipments used in the material handling to convey the material vertically. The lab model bucket elevator (Fig.34.1) consists of number of buckets/cups arranged to an endless flat belt. The belt with cups run vertically and cups are placed at a particular interval in the belt. The belt is wrapped over two pulleys, one at the top is called head pulley which receives power from an electrical motor 0.5 hp through suitable power concentration and rotates at required rpm. The belts, cups and pulleys are mounted on a steel frame. The grain from the hopper is reaches to the bottom pulley. The bucket is filled with grains during upward movement and is discharged at the top due to the centrifugal force. An outlet is provided at the top to collect the discharge grain. The capacity depends on the bucket size and seed of operation. For proper discharge of the material, the weight of the material in each cup should be equal to the centrifugal force created due to rotation.

Test procedure

- The grain loaded is the feed hopper which is kept at the bottom of the equipment and the unit is switched on.

- The rpm of the head pulley is measured using a tachometer.
- The grains are fed into the conveyer and time is noted.
- The elevator is run for s a particular time and the material collected at the outlet is weighed.

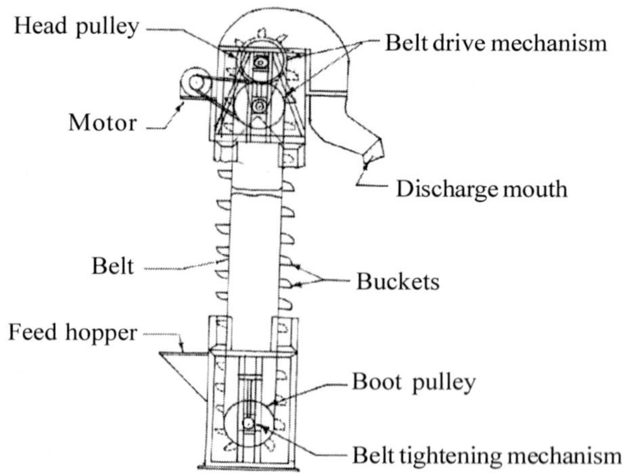

Fig. 34.1. Bucket elevator

Observation

Bulk density of the grain	(ω)	=	kg/m^3
Volume of each cup	(v_c)	=	cm^3
Spacing between two consecutive cups	(S)	=	cm
Speed of the rotation	(N)	=	rpm
Diameter of the pulley	(D)	=	m
Velocity of belt	(V)	=	m/min
Weight of the grain collected	(W_g)	=	kg
Time taken	(t)	=	h

Calculation

Speed of belt (V) $= \dfrac{\pi \times D \times N}{100}$ m / min

Theoretical capacity of bucket elevator $(T_c) = \dfrac{100}{S} \times \dfrac{Vc \times \omega}{10^6} \times V \times 60 (kg/h)$

Actial capacity of bucket elevator $(A_c) = \dfrac{W_g}{T} = kg/h$

$$\text{Conveyence efficiency } (\eta) = \frac{A_c}{T_c}$$

Result

The conveyance efficiency of the bucket elevator is __________.

35

Cost of Operation of Processing Machineries

Aim

- To determine cost of operation of processing machineries with available data.

Total operation cost

- For calculation of cost of operation of processing machineries, two types of cost viz., fixed cost and variable cost to be estimated.

Fixed cost: Fixed cost is usually calculated for every year. This includes

i) Principal for investment

ii) Depreciation

iii) Taxes

iv) Insurance

$$\text{Depreciation} = \frac{\text{Purchase value-Scrap value (sunk value)}}{\text{Life span}}$$

Variable cost: Variable cost are usually calculated every year (or) every month

i) Fuel charge/Electrical charge

ii) Repair and maintenance charge

iii) Wages and Salary for repair

iv) Lubricants, Spares etc.

Example

A seed cleaner cum grader is used for cleaning and grading of paddy. Workout the cost of cleaning and grading of 10 tone of paddy based on the data given below. Assume data wherever necessary.

Cost of the machine = Rs. 200,000/-

Life span of the machine = 25 years

Interest on investment = 12%

Housing charges = Rs. 0.8/h

Operating cost = Rs.5/h

Output = 250 kg/h

Repair and maintenance = 5% of the investment

1. Fixed cost

Assume sunk value is 5% of the machine cost

$$\text{Depreciation} = \frac{200000\text{-}(200000 \times 0.05)}{25} = \text{Rs.}7600/\text{year}$$

Annual interest on investment = 12% of principal

= 0.12×200000

= Rs. 24,000/year

Housing charges = $0.8 \times 24 \times 365$

= Rs.7008/year

Total fixed cost = depreciation + annual interest + housing rent

= $7600 + 24000 + 7008$

= **Rs. 38608/year**

2. Variable cost

Repair and maintenance = 5%of the investment

= 0.05×200000

= Rs.10000/year

Assume, the machine is running 800 h per year and its operating cost is Rs.3/h

Operating cost = 3×800

= Rs. 2400/year

Assume, consumption of electrical energy @ Rs.2 unit/h and the cost of electricity is Rs. 3/unit

Electrical cost = $2 \times 800 \times 3$

= Rs.4800/year

Total variable cost = repair charges + operating cost + electrical cost

= 10000 + 2400 + 4800

= **Rs.17200/year**

Total cost = Total fixed cost + Total variable cost

= 38608 + 17200 = Rs.55808/year

Cost per hour = 55808/800 = Rs.69.76/h

Output = 250 kg

Time taken for cleaning one tone = 1000/250 = 4 h

Cost of operation per ton = 4 × 69.76 = Rs.279.04/ton

Cost of cleaning and grading of 10 tone of paddy = 10 × 279.04 = Rs. 2790.4

36

Study of Greenhouse Construction Free Standing Gable and Quonset Type Greenhouses

Aim

- To study the construction of free standing gable and Quonset type greenhouses.

Greenhouse selection

Selection of greenhouse to be constructed is important as each type has its own merits and demerits. There are two common types of commercial greenhouses namely; (i) Freestanding (single span) and (ii) Ridge and furrow (gutter connected or multi-span). Multi-span structure permits economy in construction, heating and cooling but the single span gives flexibility as far as microclimate inside the greenhouse in concerned. The freestanding Quonset type is usually accommodates many growing situations but there height restrictions near the side walls. Another freestanding type is single gable greenhouse and its many variations. The ridge and furrow or gutter connected greenhouse are joined at the eave by a common gutter.

1. Construction of freestanding gable type greenhouse

Foundation

The foundation of greenhouse structure is critical and consulting with experience builders is highly recommended. The foundation must resist overturning and vertical pressure from structural loads. Concrete is the most appropriate material for permanent structures. If there is square foundation, size may be 60 × 60 × 60 cm and for round foundation, 30 cm diameter and 1 m depth in PCC of 1:4:8 ratios. To avoid rusting of the poles, the vertical poles may be covered to the height of 60 cm by PCC with a thickness of 5 cm.

Structural components

Greenhouse is a structure made by assembling different components and each part has specific role. It is covered with a transparent material for admitting natural light for plant growth. The main components of greenhouse are structure, covering material, temperature control systems (Fig. 36.1).

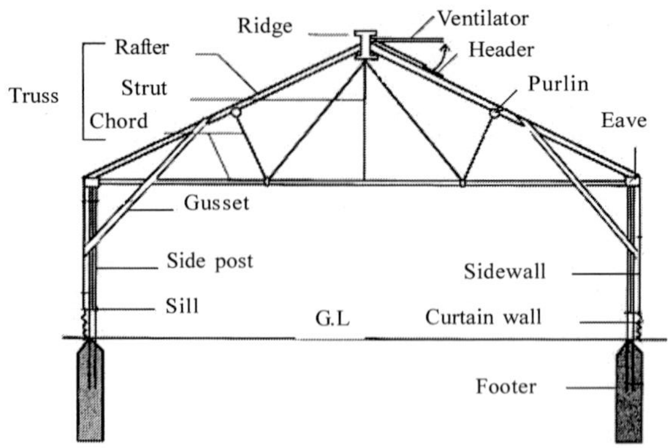

Fig. 36.1a. Structural components of a greenhouse

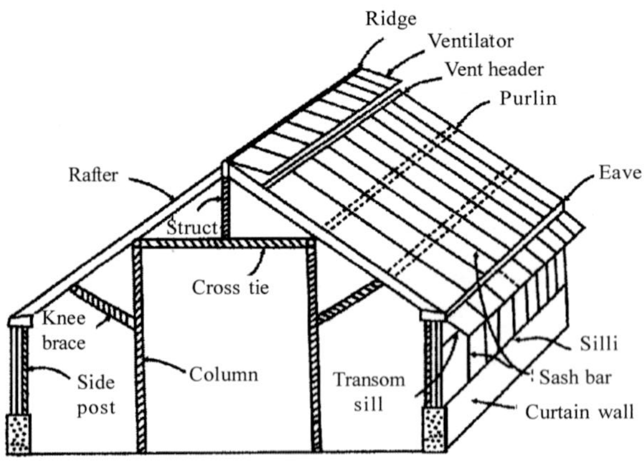

Fig. 36.1b. Structural components of a greenhouse

Side wall

It supports the trusses and bears the weight of the greenhouse. These are set in concrete footings and typically spaced 10 feet apart.

Curtain wall

The curtain wall is provided some feet of sidewall above the soil line. Usually made of some solid building materials such as poured concrete blocks, bricks, or treated lumber.

Eave

It is the top of the sides of the greenhouse and provided where the sides join the roof of greenhouse.

Truss

It is a structural component that supports the weight of the greenhouse roof. It consists of rafters, struts and chords.

Sill: It is top the curtain wall

Purlin: It runs along the length of the greenhouse and it keeps the roof trusses aligned.

Ride: Where the roofs come together at the top of the greenhouse and many greenhouses have a ridge vent(s).

Side posts and columns

These are vertical supports dictate the height of the production area. Height of them range from 0.3 to 3 m and significant consideration should be given since they directly influence efficiency.

Sash bar

It runs perpendicular to the purlins and attached with it to hold glazing in place. Sometimes built with a drip groove to catch condensation that forms on the inside of the glass panels.

2. Construction of freestanding Quonset type greenhouse

Structural members

1. Hoops
2. Foundation
3. Lateral supports
4. Polygrip assembly
5. End frame

Required materials

1. GI pipe class A (25 mm diameter)
2. GI pipe class B (15 mm diameter)
3. GI sheet (20 gauge)

4. Mild steel flat (25 x 3mm)
5. Lateral support to end frames (10 mm diameter rod)
6. Cement concrete (1:3:6 mix)
7. UV-stabilized LDPE film (single layer, 800 gauge, 5.4 m^2/kg)
8. Poly grip (channel 2000 x 3.5 x 4 cm) 20 gauge GI sheet
9. Angle 2000 x 2 x 2 cm, 20 gauge GI sheet
10. Cross connectors (6 mm diameter rod)
11. U-clamps
12. Wooden end frames (5 x 5 cm wood)
13. Nuts and bolts
14. Miscellaneous items like nails, hinges and latches as per requirement.

Preparation of materials

Hoops

Hoops are the integral part of the greenhouse frame. Hoops are formed in semicircular shape by bending the GI pipe. For bending, pipe bender is used, which can bend pipe in any desired radius. About 30 cm length on each end remains unbent which enables the ends to easily fit into the foundation pieces.

Foundation

Foundation pipes are meant to provide a firm support to the hoops and to secure the polygrip firmly. GI pipes of 25 mm (class A) and 85 cm length are used as foundation pieces. A 10 cm piece of MS flat is centrally welded to the one end of pipe and a hole of 8 mm is drilled at 10 cm distance from the other end. The flat welded end is put in a hole dug to a depth of 70 cm grouting with concrete. A 15 cm length of pipe remains above ground level to hold hoops and poly grip assembly. The foundation pipes are spaced 1.25 m apart in parallel rows and it should be seen that the top of these pipes are at the same elevation. The end frame structures should have provision for a door and installation of environmental control equipment. For smaller greenhouse (up to 50 m^2) door is provided on one side while the medium and large greenhouses may have doors on both ends (Fig. 36.2).

Lateral supports

For the lateral supports, rings of 3.5 cm diameter are made at one end and a right angle hook on the other. During assembly the ring end encircles the foundation pipe which is put on before arranging the hoops. The other end of the lateral support is hooked to the end frame. Four lateral supports are provided at the four corners.

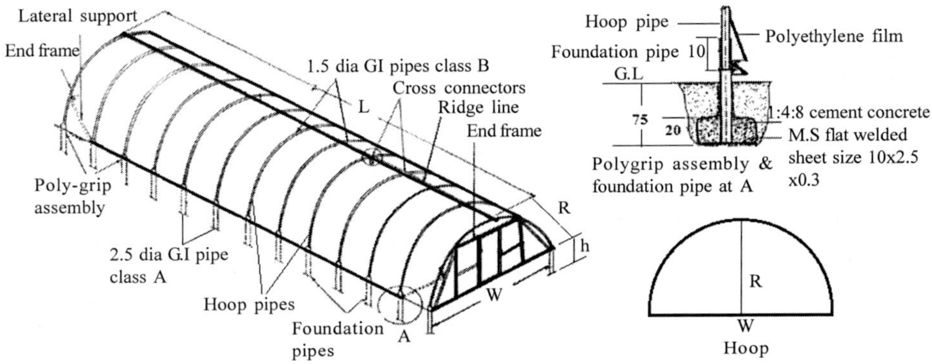

Fig. 36.2. Freestanding Quonset type greenhouse

Polygrip assembly

The function of polygrip assembly is to secure the polyethylene covering to the foundation pipes so that it can withstand the wind load and also avoiding the film puncture. The polygrip mechanism is made from 20 gauge GI sheet. Strips of 4 cm and 13 cm width with the available length are cut from the sheet. The 4 cm strip is bent centrally to form a right angle section. The 13 cm strip is bent in the shape of a channel with its edges rounded. MS rod of 6 mm diameter and 56 mm long pieces are used for positioning channel, angle and polyethylene film sandwiched in between and keeping these under pressure.

Roof

The UV-stabilized LOPE film of 150 to 200 micron thickness and width of 7 m is recommended for greenhouse glazing. The polyethylene film is stretched and these MS rod pieces are put at a distance of 50 cm, holding the right angle strip against the channel along the whole length of the greenhouse on both sides. The film is secured to both sides along the length using the poly grip mechanism and to end frames with wood nails. A 15 mm diameter GI pipe fastened at the ridge line maintains the hoops spacing and they are firmly secured to the hoops by the ridge line grip mechanism, which is fabricated from MS rods, looped over the crossing of the ridge and tightened with a bolt.

Procedure of construction

A 4×20 m rectangular area is marked on the site, preferably orienting the longer dimension in east-west direction. This rectangle will act as the floor plan of the greenhouse. Make sure that the adjacent sides are at right angles, which can be ensured by maintaining the two diagonals of the rectangle equal. Mark four points on the four comers of the rectangle. Start from one comer point and move along the length of the marked rectangle, marking a point every 1.25 m distance until reaching the other comer.

The procedure is repeated on the other side of the rectangle. Dig 10 cm diameter holes to 70 cm depth on all marked points with the help of bucket auger or a crow bar. This way a total of 34 holes on both the parallel sides of the greenhouse floor are obtained. Splice poly grip sections of available lengths formed according to the drawing into two 20 m lengths. Fix the prefabricated poly grip channels to the foundation pipes on 1.25 m spacing with the help of 6 mm diameter bolts.

Set these assemblies on temporary supports between the holes with the foundation pipes hanging vertically in the holes and the tops straight and at constant elevation. Pour cement concrete mix of 1:3:6 around foundation pipes such that the lower 15 to 20 cm pipe ends are covered in concrete. The concrete is compacted around the foundation pipes with the help of the crow bar and is allowed to cure for 2 to 3 days.

After curing, fill the soil around the foundation pipes to the ground level and compact it well. Mark the position of legs and dig holes for the fixing of legs. Now install both the end frames vertically and duly compact soil on the legs. Put the ringside of lateral support members on adjacent foundation pipe to the comer, and other side is hooked to the end frame.

Put all the hoops in foundation pipes such that the straight portion of hoop is inserted into the foundation and rests on the bolt used for fixing of the polygrip channel. Put 20 m long pipe at the ridge line of the hoops. Use cross connectors on the ridge line pipe such that one half of it remains on the one side of the hoop and the other half on the other side. Put two bolts of 6 mm in the holes provided in the ends of the cross connector.

Repeat the procedure for joining all the hoops with ridge line pipe. While fixing cross connectors, the distance between the hoops or cross connectors should be maintained 1.25 m centre to centre. Spread polyethylene film over the structure from one end to the other without wrinkles.

Place polyethylene film between the polygrip channel and right angle strip and secure them under pressure with the help of the iron rods. The film is stretched gently and fixed on the other parallel side by poly grip. On the other two remaining ends, polyethylene is nailed to the end frames using wooden battens and nails.

37

Design of Greenhouse Heating System

Aim

- To study on design of greenhouse heating system
- To determine the heat requirement during winter season

Theory

Heat required for heating greenhouses is based on the quantity of heat lost through different structural components of greenhouses per unit time. Shapes of greenhouse (A-frame and quonset) and structural components (roof, wall, gable, curtain wall, curved covering and end) are mainly influencing the calculation of heat requirement. The amount of heat lost from a greenhouse depends on glazing materials, exposed surface area of the house, inside desired temperature, outside existing temperature, wind speed, and condition of house (old, new, tight or loose). The outside temperature should be based on the lowest average expected temperature during the heating season.

Heat requirement calculation

Simple and quick approach for calculating the heat requirements is based on heat transfer occurring across greenhouse defined by the overall heat transfer equation.

The amount of heat lost from a greenhouse depends on

i) Glazing materials

ii) Exposed surface area of the house

iii) Inside desired temperature

iv) Outside existing temperature

v) Wind speed and

vi) Condition of house (old, new, tight or loose).

The outside temperature should be based on the lowest average expected temperature during the heating season.

$$Q = U \times A \times (T_i - T_o)$$

Where,

Q = total amount of heat transferred the greenhouse (Watts)

U = Overall heat transfer coefficient for the material ($W/m^2/^oC$)

A = Exposed surface area of the structural material (m^2)

T_i = Inside desired temperature (°C)

T_o = Outside temperature (°C)

Overall heat transfer equation is accounts for the overall heat loss through the greenhouse structural material at general conditions, and will not account for losses due to excessive wind velocities, loose fittings, or ventilation. To account for these losses, an additional 10% can be added to the calculated total heat. The overall heat transfer coefficient (U) value should be varying with structural materials (glazing and curtain wall), and weather conditions. Exposed heat transfer surface area (A) can be calculated from the geometry of greenhouse.

Table 1. U-values of typical greenhouse glazing materials (ASAE Standards, 2002).

S. No	Type of materials	U ($W/m^2/^oC$)
1.	Brick wall (varies with material, thickness, insulation, etc)	2.5 - 3.7
2.	Glass double pane	3
3.	Acrylic and polycarbonate sheets	3.5
4.	FRP	5.7
5.	IR-absorbing and PVC film	5.7
6.	UV-stabilized polyethylene films	6.3
7.	Glass single pane	6.3

Solved example 1

Calculate the heat requirement for a metal framed glass greenhouse of 9.8 m wide by 30.5 m long and wall height of 1.8 m. The central height at the gable end above the wall is 2.74 m. Take the overall heat transfer coefficient U for glass is 6.53 $W/m^2\,^oC$. The existing outside temperature is -17.8°C and desired inside temperature would be 15.5°C (Boodley, 1996).

Solution

Length = 30.5 m Width = 9.8 m

Height = 1.8 m Central height = 2.74 m

U = 6.53 W/m^2 °C T_o = - 17.8 °C; T_i = 15.5°C

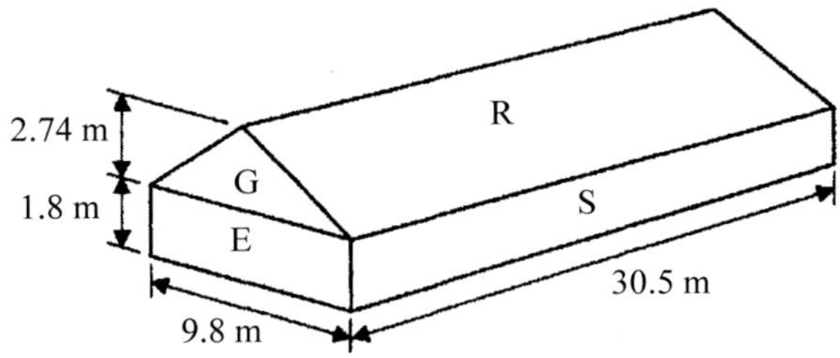

Fig.37.1. A-metal framed glass greenhouse

Area of side walls (S) = 30.5 × 1.8 × 2 (two side walls) = 109.8 m^2

Area of end walls (E) = 9.8 × 1.8 × 2 (two end walls) = 35.28 m^2

Area of gable ends (G) based on area of triangle

= ½ × 9.8 × 2.74 × 2 (two gable ends) = 26.85 m^2

The inclined length of the roof (rafter length) can be calculated from the geometry of right angled triangle from the sketch as:

Inclined length of roof = $\sqrt{(9.8/2)^2 + 2.74^2} = 5.61$ m

Area of roofs (R) based on their inclined length and length of greenhouse

= 5.61 × 30.5 × 2 (two roof slopes) = 342.21 m^2

Hence, the total exposed area of heat transfer (A)

= 109.8 + 35.28 + 26.85 + 342.21 = 514.14 m^2

Substituting in overall heat transfer equation

Heat requirement (Q) = U × A × (T_i - T_o)

= 6.53 × 514.14 × [15.5 - (-17.8)] = 111799 w = 112 kW

Heaters should be selected with a net effective heating capacity of 112 kW. For greenhouses of 30.5 m length or longer, efficient heat distribution is achieved by two small heaters with total heating capacity matching the calculated heat loss.

Solved example 2

A gable type greenhouse is constructed with 60 m long, 15 m wide and 1.5 high. The walls of greenhouse are made up of UV-stabilized polyethylene films. The central height at the gable end above the wall is 2.6 m. The polyethylene wall is established on the curtain wall of 0.5 m height and is made up of brick wall. The existing outside temperature is -10°C and required inside temperature would be

16°C. Calculate the heat requirement. Overall heat transfer coefficients for polyethylene film and brick wall are 6.3 and 3.2 $W/m^2 °C$ respectively.

Solution

Length = 60 m Width = 15 m

Height = 1.5 m Central height = 2.6 m

T_o = - 10°C T_i = 16°C

U_p = 6.3 $W/m^2 °C$ U_b = 3.2 $W/m^2 °C$

For curtain wall

Area of curtain wall (A_c) = (60 × 0.5 × 2) + (15 × 0.5 × 2) = 900 m^2

Heat requirement for curtain wall (Q_c) = $U_c \times A_c \times (T_i - T_o)$

= 3.2 × 900 × [16 – (-10)] = 74880 W = 74.88 kW

For polyethylene wall

Area of side walls (S) = 60 × 1.5 × 2 (two side walls) = 180 m^2

Area of end walls (E) = 15 × 1.5 × 2 (two end walls) = 45 m^2

Area of gable ends (G) based on area of triangle

= ½ × 15 × 2.6 × 2 (two gable ends) = 39 m^2

The inclined length of the roof (rafter length) can be calculated from the geometry of right angled triangle from the sketch as

Inclined length of roof = $\sqrt{(15/2)^2 + 2.6^2} = 7.94$ m

Area of roofs (R) based on their inclined length and length of greenhouse

= 7.94 × 60 × 2 (two roof slopes) = 952.8 m^2

Total area of polyethylene wall and roof (A_p)

= 180 + 45 + 39 + 952.8 = 1216.8 m^2

Heat requirement for polyethylene wall and roof (Q_p) = $U_p \times A_p \times (T_i - T_o)$

= 6.3 × 1216.8 × [16 – (-10)] = 199311.8 W = 199.32 kW

Total heat requirement (Q) = $Q_c + Q_p$ = 74.88 + 199.32 H” 275 kW.

Solved example 3

Calculate the heat requirement for a double layered polyethylene arch roof greenhouse of 9.8 m wide by 30.5 m long and wall height of 1.8 m. The central height at the circular arc hoop above the wall is 2.74 m. Assume overall heat transfer coefficient for double layer polyethylene as 4.543 W/m²/°C. The existing outside temperature is -17.8°C and desired inside temperature would be 15.5°C.

Solution

Length = 30.5 m	Wall height = 1.8 m
Width = 9.8 m	Central height = 2.74 m
U = 4.543 W/m²/°C	T_o = -17.8 C & T_i = 15.5°C

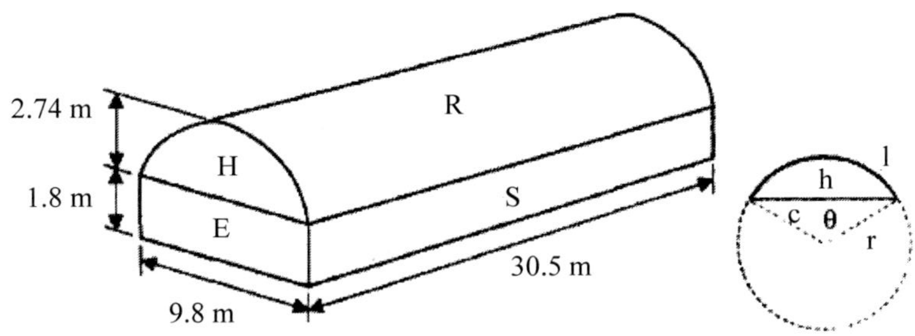

Fig. 37.2. Double layered polyethylene arch roof greenhouse

Area of walls

Area of end walls (E) = 9.8 × 1.8 × 2 (two end walls) = 35.28 m²

Area of side walls (S) = 30.5 × 1.8 × 2 (two side walls) = 109.8 m²

Area of arch section

Assume, arch section as a portion of circular arc

c = chord length (greenhouse width as chord length)

h = height of arch at the middle (maximum height of the arc from the cord) above the wall

l = Length of the arc

r = radius of the arc

θ = subtended angle in radians

S = enclosed area of the arc

From standard geometrical formula of segment of circle

$$r = \frac{c^2 + 4h^2}{8h}$$

$$\theta = 2 \times \arcsin\left(\frac{c}{2r}\right)$$

$$l = r \times \theta$$

$$S = \frac{r^2}{2}(\theta - \sin\theta)$$

Substituting the values in the above equations:

$$r = \frac{9.8^2 + (4 \times 2.74^2)}{8 \times 2.74} = 5.75 \text{ m}$$

$$\theta = 2 \times \arcsin\left(\frac{9.8}{2 \times 5.75}\right) = 2.039 \text{ radian}$$

$$l = 5.75 \times 2.039 = 11.73 \text{ m}$$

$$S = \frac{5.75^2}{2}[2.039 - \sin(2.039)] = 18.96 \text{ m}^2$$

Area of arch portions (H) based on the contained area (S)

= 18.96 × 2 (two arch ends) = 37.92 m^2

Area of arch roof (R) based on the are length (l)

= 30.5 × 11.73 × 1 (whole arch length) = 357.77 m^2

Total exposed area of heat transfer (A)

= 35.28 + 109.8 + 37.92 + 357.77 m2 = 540.77 m^2

Heat requirement (Q)

= $U_p \times A_p \times (T_i - T_o)$

= 4.543 × 540.77 × [15.5 – (-17.8)] = 81809 W = 82 kW

Hence, a heater with net effective capacity of 82 kW should be chosen.

Arch roof geometry will be more simplified if the arch is a semi-circle. Quonset type greenhouses can be visualized as direct segment of circles without the straight wall portion, as found in arch type houses.

Solved example 4

Calculate the heat requirement for increase inside temperature of double layered polyethylene Quonset type greenhouse up to 15°C. The central height of the hoop from ground level is 6 m. The length of house is 85 m. The existing outside temperature is 3°C. The overall heat transfer coefficient for double layer polyethylene is 4.5 W/m^2°C.

Solution

Length of greenhouse (L) = 85 m

Central height of the hoop from ground level (r) = 6 m

Outside temperature (T_o) = 3°C

Inside desired temperature (T_i) = 15°C

Heat transfer coefficient for double layer polyethylene (U) = 4.5 W/m^2°C

Heat requirement (Q_g):

Arch covering width (d) = $\pi \times r = 3.14 \times 6 = 18.84$ m

Double layer plastic roof area (A_r) = $d \times L = 18.84 \times 85 = 1601.4$ m^2

Double layer plastic end walls area (A_{ew}) = $\pi \times r^2 = 3.14 \times 6^2 = 113.04$ m^2

Total area of the double layer plastic (A) = 1601.4 + 113.04 = 1714.44 m^2

Heat requirement (Q_g) = $U \times A \times (T_i - T_o)$

= $4.5 \times 1714.44 \times (15 - 3) = 92579.76$ W = 92.6 kW

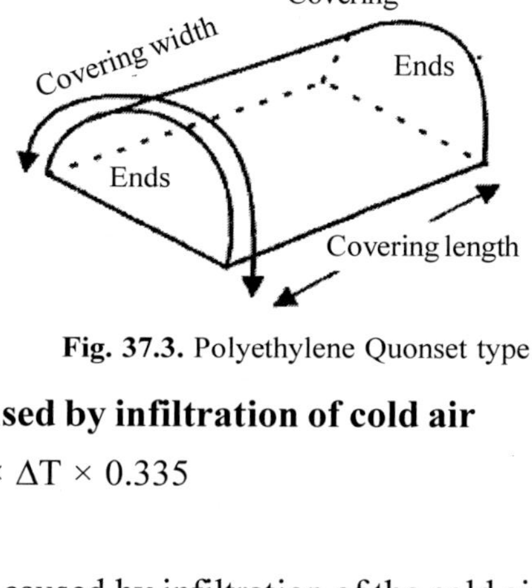

Fig. 37.3. Polyethylene Quonset type greenhouse

Heat loss caused by infiltration of cold air

$Q_i = C_a \times V_g \times \Delta T \times 0.335$

Where,

Q_i = Heat loss caused by infiltration of the cold air (W/m^2°C)

C_a = Number of air changes per hour (0.25)

V_g = Volume of the greenhouse (m^3)

ΔT = Temperature difference between inside and outside the greenhouse (°C)

$$V_g = \frac{\pi \times r^2}{2} \times L = \frac{3.14 \times 6^2 \times 85}{2} = 4819.5 m^3$$

$Q_i = 0.25 \times 4819.5 \times (15 - 3) \times 0.335 = 4843.598$ W = 4.84 kW

$Q = Q_g + Q_i$ = 92.6 + 4.84 H” 98 kW

38

Design of Greenhouse Cooling System

Aim

- To design of greenhouse cooling system
- To determine the rate of air exchange in an active summer and winter cooling system.

Theory

Elevation factor (F_{elev})

Elevation factor is directly proportional to the elevation. The values of elevation factors, used to correct the rate of air removal for a particular elevation. The rate of air removal from the greenhouse increase as the elevation of the greenhouse site increases.

Light factor (F_{light})

The rate of air removal is also depends upon the light intensity in the greenhouse. As light intensity increases, the heat input from the sun increases, requiring a greater rate of air removal from the greenhouse. Hence light factor is directly related to the light intensity that expressed by klux. In general, an intensity of 53.8 klux is accepted as a desirable level for crops.

Temperature factor (F_{temp})

Solar energy warms the air as it passes from the pad to the exhaust fans. Usually, a 4°C rise in temperature is tolerated across the greenhouse. If it becomes important to have uniform temperature across the greenhouse, it will be necessary to raise the velocity of the air movement through the greenhouse. To maintain less temperature difference across the greenhouse more air is to be circulated. Hence, the temperature factor (F_{temp}) increases as the difference in temperature across pad to fan decreases.

Velocity factor (F_{vel})

The pad and fans should be placed on opposite walls, either end walls or side walls of the greenhouse and the distance between them is important. A distance of 30 to 61 m is the best. To achieve a given degree of cooling, more amount of air is required when pad to fan distance is less and vice versa. So the velocity of incoming air is to be modified accordingly. As per the guidelines of the National Greenhouse Manufacturers Association-2004, for ventilating and cooling greenhouses, rate of air removal from one square meter greenhouse floor is 2.5 m^3/min. This applies to a greenhouse under 305 m in elevation, with an interior light intensity not exceeding 53.8 klux and an air temperature rise of 4°C from pad to fans.

Light intensity

Light output is expressed in lumens and light intensity is measured in terms of lumens per square meter or lux. Total energy of all the light given off is termed the "luminous flux"

Air exchange rate (Q_{std})

Rate of air exchange is measured in cubic meters per minute (m^3/min or cmm). The rate of air removal required for a greenhouse can be calculated under standard conditions (Q_{std}) using the following equation:

$$Q_{std} = L \times W \times 2.5$$

Where, L is the length (m) and W is the width (m) of greenhouse.

Adjusted air removal rate (Q_{adj})

The standard rate of air removal is adjusted by multiplying it by F_{house}, or F_{vel}. The F_{house} is calculated from the following equation:

$$F_{house} = F_{elev} \times F_{light} \times F_{temp}$$

Thus, the final adjusted (Q_{adj}) capacity of the exhaust fans must be:

$$Q_{adj} = Q_{std} \times F_{house}$$

Fan capacity

Fans should not be spaced more than 7.6 m apart. The required capacity or size of each fan can be determined by following relationship:

$$\text{Fan capacity} = \frac{Q_{adj}}{\text{number of fans}}$$

Pad size

The excelsior (wood fibre) pads of 2.5 to 4 cm thick are generally used. These are replaced annually and they support an airflow rate of 45 $m^3/min/m^2$. Cross-fluted cellulose pads come in units of 30 cm wide and are 5, 10, 15 or 30 cm thick.

Pad area

As per recommendation of the NGMA (2004), one square meter is required for each 75 m^3/min of fan capacity.

$$A_p = \frac{F_c}{75} \times N_f \text{ or } A_p = \frac{V_0}{V_i}$$

Where, A_p is the pad area (m^2), F_c is the fan capacity; N_f is the number of fans; V_o is the volume of air that must be removed from the greenhouse in unit time, and V_i is the volume of air that can be moved through a square meter of pad.

Pad height

The pad should cover the width of wall in which it is to be installed. Height of pad can be determined by the following relationship:

Pad height = Total pad area/pad width

Pump capacity

The pump capacity is based on a standard air flow rate of 6.2 lpm/m for 10 cm thick pad multiplied by the length of the pad.

Pump capacity = Standard air flow rate × length of the pad

Sump size

The sump size is based on a standard volume rate of 30.5 lit/m^2 for 10 cm thick pad multiplied by the actual pad area.

Sump size = Standard volume rate × Actual pad area

Active summer cooling system

Solved example 1

Design a summer evaporative cooling system with the following requirements. Consider a single greenhouse 15 m wide and 30 m long located at an elevation of 915 m. The greenhouse has a moderate coat of shading compound on it, providing the maximum light intensity of 53.8 klux. The temperature rise of 4°C can be tolerated from pad to fans. Use a 10 cm thick cross-fluted cellulose cooling system for the greenhouse. Assume, $F_{elev} = 1.12$, $F_{light} = 1.0$ and $F_{temp} = 1.0$ (Nelson, 2003)

Solution

Length of greenhouse (L) = 30 m   Width of greenhouse (W) = 15 m

Maximum light intensity = 53.8 klux   Elevation = 915 m

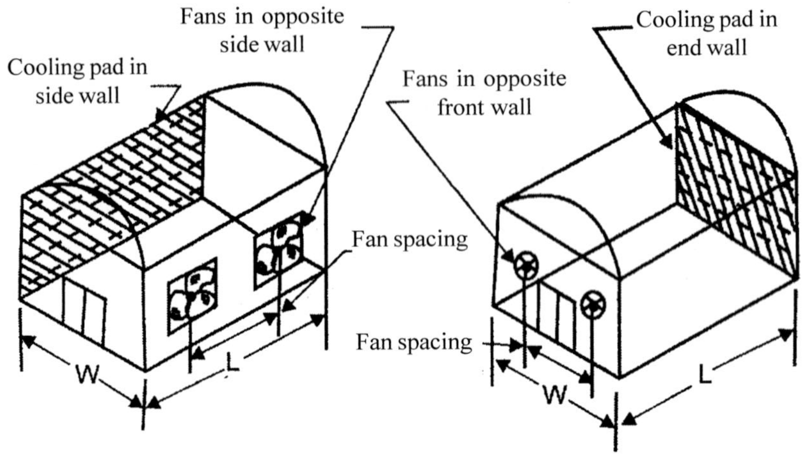

Fig.38.1. Arrangement of fan and pad in active summer cooling system

Quantity of air to be removed per minute under standard conditions:

$Q_{std} = L \times W \times 2.5\ m^3/min/m^2$

$= 30 \times 15 \times 2.5 = 1125\ m^3/min$

Greenhouse factor:

$F_{house} = F_{elev} \times F_{light} \times F_{temp}$

$= 1.12 \times 1 \times 1 = 1.12$

Adjusted volume of air to be expelled from the greenhouse each minute:

$Q_{adj} = Q_{std} \times F_{house}$

$= 1125 \times 1.12 = 1260\ m^3/min$

Number of fans required: Fans should not be over 7.6 m apart

= Width or Length of house/7.6

= 15/7.6 = 2 fans

Fan capacity or size

$$\text{Fan capacity} = \frac{Q_{adj}}{\text{number of fans}}$$

= 630 m^3/min/fan

From Table, 2 numbers of 137 m^3 fans with 1 hp motors would be selected.

Pad area

As per NGMA (2004), the fan capacity of 75 m^3/min is required per square meter pad

$$A_P = \frac{F_c}{75} \times N_f = \frac{630}{75} \times 2 = 16.8\,m^2$$

Pad height

The pad should cover the width of wall in which it is to be installed (15 m width).

Pad height = Total pad area/pad width = 16.8/15 = 1.1 m

Theoretical pad height is 1.1 m but a pad height of 1.22 m is available in the market

Actual pad area = 15 m × 1.22 m = 18.3 m^2

Pump capacity

The pump capacity is based on a standard air flow rate of 6.2 lpm/m for 10 cm thick pad multiplied by the length of the pad.

Pump capacity = 6.2 lpm/m × 15 m = 93 lpm

Sump size

The sump size is based on a standard volume rate of 30.5 lit/m^2 for 10 cm thick pad multiplied by the actual pad area.

Sump volume = standard volume rate × actual pad area = 30.5 × 18.3 = 558 lit

Active winter cooling system

During winter, air temperature inside the greenhouse is more than the outside air temperature. Hence actual winter cooling is done by mixing the outside ambient air with the help of convection tubes. In active winter cooling systems, under standard conditions, a volume of 0.61 m^3/min of air should be removed from the greenhouse for each square meter of floor area. Cold air must be introduced into the greenhouse at a greater rate if desire to reduce the inside air temperature. In active winter cooling, the winter factor (F_{winter}) based on temperature difference between inside and outside air vary inversely with the required temperature difference.

Solved example 2

Design a winter convection tube cooling system for the following requirements. Consider a single greenhouse 15 m wide and 30 m long located at an elevation of 915 m. The greenhouse has a moderate coat of shading compound on it, providing the maximum light intensity of 53.8 klux, and the desired interior-to-exterior temperature difference of 8°C. Assume, F_{winter} is 1.0 and F_{light} is 1.12 (Nelson, 2003)

Solution

Area of greenhouse (A) = L × W = 30 × 15 = 450 m^2

Capacity of the exhaust fan under standard conditions:

$Q_{std} = A \times 0.61\ m^3/min/m^2 = 450 \times 0.61 = 275\ m^3/min$

Adjusted exhaust fan capacity:

$Q_{adj} = Q_{std} \times F_{winter} \times F_{light} = 275 \times 1.0 \times 1.12 = 308\ m^3/min$

The capacity of pressurizing fan in the inlet end of the convection tube is equal to the capacity of exhaust fan. Two pressurizing fans needed in this case of a combined capacity of 308 m^3/min, which is equal to the exhaust fan capacity.

Size and number of air distribution tubes for greenhouse having 15 m width and 30 m length can be selected from Table 11.7. Diameter of the tube is 61 cm and the number of tubes is 2.

Cross-sectional area of tube

$$A = \frac{\pi d^2}{4} = \frac{\pi \times 61^2}{4} = 2922 cm^2$$

Combined area of all the holes in a tube:

In general, total area of all the holes in a single tube should be 1.5 to 2 times the cross sectional area of the tube.

Combined area of all the holes in a tube = A × 1.5 to 2 = 4384 to 5845 cm^2

Number of holes

The diameter of the individual holes along the side of distribution tubes and the distance between them must be decided. If tube length increases the distance between holes increase to maintain a reasonable diameter hole. A hole spacing of 61 to 122 cm is common. Considering the combined area of holes as 5845 cm^2 and length of tube as 30 m (3000 cm), the hole spacing is 61 cm.

Number of holes per side of the tube

= Length of tube / Hole spacing

= 3000/61 = 49

Number of holes in one tube

= 2 × 49 = 98 (Holes are to be provide on both sides of the tube)

Area of the hole

A = Cross sectional area of tube / Number of holes

= 5845/98 = 59.64 cm^2

Diameter of the hole

$$d=\sqrt{\frac{4\times A}{\pi}}=\sqrt{\frac{4\times 59.64}{\pi}}=8.71\text{cm}$$

Exercise 1: Design a summer evaporative cooling system for the following requirements. Greenhouse is 20 m wide and 40 m long located at an elevation 600 m, experiencing a maximum light intensity of 70.0 klux, and a tolerable temperature difference from pad-to-fan is 5°C. A cross fluted cellulose of 15 cm thick cooling system was used.

Exercise 2: Design a winter convection tube cooling system for the following requirements. Size of the greenhouse is 20 m width and 40 m length located at an elevation 600 m, experiencing a maximum light intensity of 70.0 klux, and the desired interior-to exterior temperature difference is 7°C.

Exercise 3: A gable type greenhouse is constructed with 115 m long, 18 m wide and 1.8 high. The walls of greenhouse are made up of UV-stabilized polyethylene films. The central height at the gable end above the wall is 2.4 m. The polyethylene wall is established on the curtain wall of 0.7 m height and is made up of brick wall. The existing outside temperature is -5°C and required inside temperature would be 15°C. Calculate the heat requirement. Overall heat transfer coefficients for polyethylene film and brick wall are 6.35 and 3.25 W/m^2 C respectively.

Exercise 4: Calculate the heat requirement for increase inside temperature of double layered polyethylene Quonset type greenhouse up to 18 C. The central height of the hoop from ground level is 8 m. The length of house is 125 m. The existing outside temperature is -7°C. The overall heat transfer coefficient for double layer polyethylene is 4.75 W/m^2 C.

39

Study of Chain Survey Tools and Equipment

Aim

- To study the different materials used for chain survey

Theory

Chain surveying is the method of land surveying in which only linear measurements are made. This method of surveying quit simple and it requires only a chain, a tape, few ranging rods and few arrows. It is used for securing data for exact description, marking the boundaries of land and preparing the maps of the area to show various details. The term details mean a natural or man-made feature at the ground surface.

Materials for chain survey

- Materials used for measuring distances

1. Chain
2. Tape

- Materials used for marking survey stations

1. Ranging rods
2. Offset rods
3. Laths and whites
4. Pegs

- Materials used for setting right angles

1. Cross staff
2. Optical square

- Other materials

1. Arrows or chain pins
2. Plumb bob

Metric chain

Metric chain is commonly used for land survey and commonly used for measurement of horizontal distances. The chain is composed of number of links, made up of galvanized mild steel wire of 4 mm diameter. The ends of each link bent into a loop and connected together by means of three small connecting rings (Fig.39.1a). The rings consist of outer two rings of oval shape whereas the central ring is circular. These rings which offered flexibility to the chain and make less liable to become kinked. The joints of the links are usually open, but in the best chains they are welded so as to render the chain less liable to stretching. The ends of the chain provided with brass handles for dragging the chain on the ground, each with a swivel joint, so that the chain can be turned round without twisting. The handles have swivel joints to provide flexibility so that it is possible to swivel the handle round the eyebolt without it's twisting.

The length of a link is measured from the centre of the connecting central ring to the centre of the next connecting central ring, and it is 20 cm for all links except the end links. The end links include the handles (Fig.39.1b). Metallic tags or indicators of distinctive pattern are fixed at various distinctive points of the chain to felicitate quick reading of fraction of a chain in surveying measurements. To facilitate holding the chain pin at end of the chain, a semi-circular groove of the same radius as that of the chain pin is cut on the outside surface of the handle. The end links are smaller than the other links because the length of the brass handles is included in 20 cm length. To enable the reading of fraction of a chain length without difficulty, small brass rings are provided at every meter length and brass tallies are provided at every 5 meter length in the 20 m and 30 m long chains. The overall length of the chain is taken as the distance between the outsides of the two handles when the chain is fully stretched.

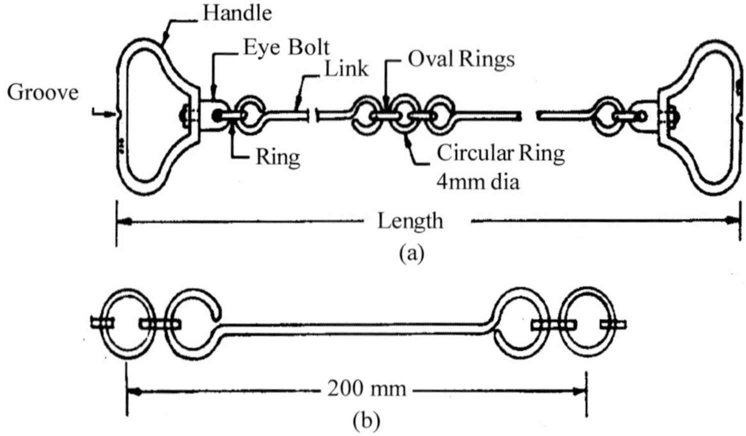

Fig.39.1. Metric chain

Ranging rods

Ranging rods are used to locate intermediate points such that these points lie on the straight line joining the end stations. Ranging is essential before starting the measurement of a line when length of line is more than a chain length. Ranging rods (Fig.39.2a) are made of either well-seasoned straight timber or mild steel tubular section. They are circular in cross-section, with 30 mm nominal diameter. The bottom end is fitted with a hollow, cast iron shoe or steel sheet shoe so that it can be easily inserted into the ground. There are two lengths of ranging rods, one is 2 m length with 10 bands of 20 cm each and other is 3 m length with 15 bands of 20 cm each. The bands are painted alternatively either black and white or red and white to make the rod visible even from a long distance.

Offset rods

An offset rod is used to setting out offset lines at right angles to the survey line, and for measuring small offsets. The offset rod is similar to ranging rods, but has two short slots passing through the center of the section of the rod (Fig.39.2b). The slots are made at right angles to each other at about the eye level of the person when the rod is held vertically. The length is 3 m.

Arrows or chain pins

Arrows or chain pins (Fig.39.3) are used to mark the positions of the ends of the chain on the ground and record the number of chain lengths laid down when measuring a long line. These are made of good quality, hardened and tempered steel wire of 4 mm diameter and the length varies from 25 to 50 cm. The arrow has a loop at one end, whereas the other end is pointed but slightly rounded. Generally 10 arrows are supplied with a chain.

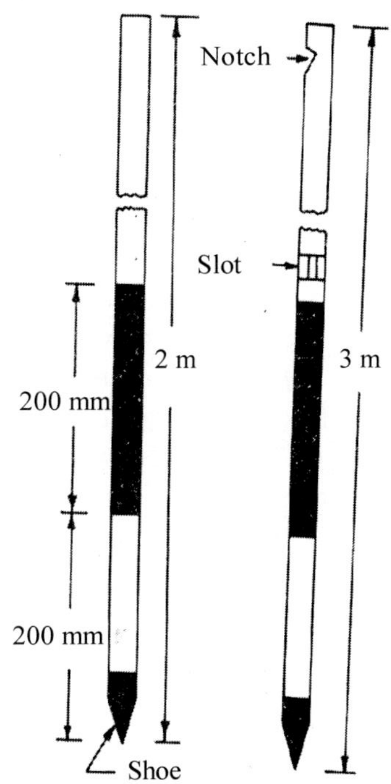

Fig.39.2. (a) Ranging rod
(b) Offset rod

Fig.39.3. Chain pin

Pegs

Wooden pegs are used to mark the positions of the survey stations or end points of a survey line. They are made of hard timber and are tapered at one end (Fig.39.4). They are usually, 2.5 cm square and 15 cm long, but in soft ground, pegs 40 to 60 cm long and 4 to 5 cm square suitable.

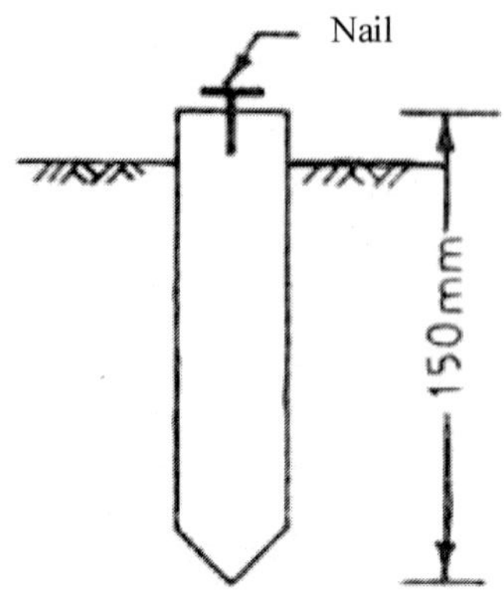

Fig.39.4. Peg

Cross-staff

Cross-staff is used for finding the foot of the perpendicular from a given point to a line, and setting out a right angle at a given point on a line. The simplest form of cross staff is the open cross staff. It consists of two parts (i) the head and (ii) the leg. The head consists of four metal arms with vertical slits. The arms are rigidly fixed in such a manner so that the center of one pair of arms forming a straight line makes right angle with the other pair of arms. In one line, one of the slits is narrower than the other. One horse hair is fixed at the center of the wider slit. The object is sighted from the narrow slit in line with the hair. The cross staff is mounted on 25 mm diameter, about 1.5 m long pole for fixing on the ground (Fig.39.5).

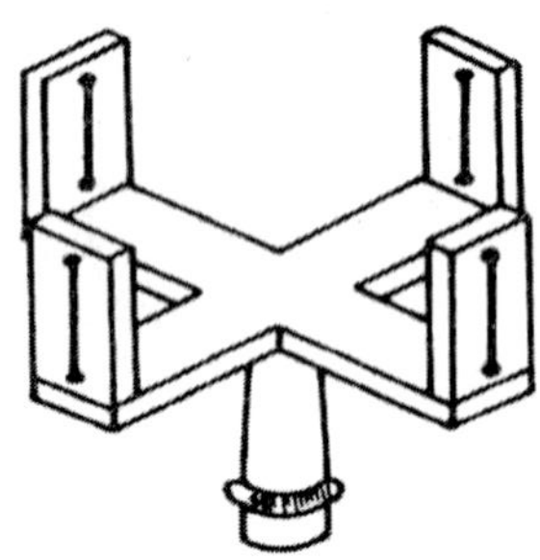

Fig.39.5. Open cross-staff

For laying out a right angle at a point on the chain line, the cross staff is held vertically on the supporting pole at the given point. Ranging rod is fixed on the chain line on either side of the cross staff and sighted through the slit and horse hair. The cross staff is turned till the ranging rod is visible. At this time, one sight through the other pair of slits and another person fixes a ranging rod in this line of sight. Foot of the cross staff joined with the ranging rod gives perpendicular line with the chain line.

40

Measurement of Horizontal Distance Using Metric Chain

Aim

- To measure horizontal distance on the level ground using chain

Materials required

- Chain
- Ranging rods
- Chain pins
- Field note book

Unfolding the chain

- Remove the leather strap from the folded chain. Take both the handles in the left hand, and hold the remainder of the chain in the right hand.
- Holding both the handles in the left hand, throw the remainder of the chain forward on the ground.
- Lay the handles on the ground so that the two halves of the chain lie side by side.
- Leaving one handle on the ground, pick up the other handle and walk forward until the chain is fully extended.

Folding the chain

- Bring the two handles together on the ground so that the chain lies doubled up. Pull the chain from the middle point to make the two halves straight.
- With right hand, pick up the two links on either side of the middle tally and hold them side by side slightly obliquely, in the left hand.
- With the right hand, grasp the next two links and hold them together. Lay these two links at a slight angle across the first two links in the left hand.

- Continue the procedure until the whole chain has been rolled up. Fasten the leather strap firmly around the waist of the links so formed.

Procedure for chaining a line

- In an ideal chaining, at least 3 persons are required. The person at the forward end is called the leader. The person at the rear end is called the follower. The person directs the work and keeps the record of the measurement in a field book is called the surveyor.
- The chain is unfolded and stretched then it should be examined before starting the measurement.
- The follower stands behind the ranging rod at starting point (A) and holds firmly the zero end of the chain in conduct with the peg at point A.
- The leader takes arrows and ranging rod and the forward handle of the chain, and proceeds in the forward direction. He drags the chain after him.
- As soon as reaches a particular point (B), holds the ranging rod vertically in contact with the outside of the handle on the line AB then faces the follower for correct ranging.
- The follower directs the leader using the code of signals to move the ranging rod to right or left as required until it is exactly on the line AB.
- The follower now holds the rear handle of the chain exactly over the peg at point A.
- The leader takes the forward handle in his both hands, and straightens the chain by jerking. After the chain has been straightened, the leader holds an arrow against the end of the handle.
- If the ground is hard, a cross is scratched on the ground with a chalk and an arrow is placed horizontally at point.
- The leader then picks up the ranging rod and the remaining arrows and proceeds in the forward direction with the handle of the chain.
- The follower carrying the ranging rod and rear handle of the chain also walks in the forward direction.
- When the follower reaches the arrow fixed, gives indication to the leader and he stops moving forward and turns round towards the follower.
- The follower then places the ranging rod vertically behind the arrow already fixed and directs the leader to place his ranging rod on the line, as above.
- The follower holds the rear handle against the first arrow and the leader stretches the chain straight in the line and inserts the second arrow into the ground, and proceeds further.

- The follower then removes the first arrow and picks up the ranging rod and takes the rear handle of the chain in one hand and walks forward.
- The whole process is repeated until the end B of the line is reached.
- If the distance is greater than 10 chain lengths, arrows are transferred from the follower to the leader.
- At end of the line, there is generally a fractional part of the chain. To measure the fractional part, the leader drags and stretches the chain straight beyond the end station B while the follower holds the rear handle against the last arrow.
- The tag just before the point B is noted and the number of 1 m rings and 20 cm links is counted. After the completion of the chaining operation, the chain should be folded.

Procedure for ranging a line

- When the distance is greater than the chain length and end station is not clearly visible, it is necessary to establish intermediate points to maintain the direction.
- The process of establishing the intermediate points on a long survey line between the terminal points is called ranging or ranging out.
- Ranging should be done prior to chaining. Ranging is usually done by using ranging rods.
- When the end survey stations are visible from one another, intermediate ranging rods are placed in line by direct observation, the process is known as direct ranging.
- Let, A and B be first and end survey station respectively and are visible from one another (Fig.40.1). Erect the ranging rods vertically both the stations A and B.
- For erecting the ranging rod, hold it firmly in the right hand such that its lower end is about 45 cm above the ground and then drive it firmly into the ground by a quick downward blow.
- After erecting the ranging rods at A and B, locate a point C lies in the straight line joining AB at a distance not greater than one chain length from the starting station A.
- The surveyor stands behind the ranging rod at A at a distance of about half meter.
- Another person (assistant) stands and holds a ranging rod vertically at point C.

- The assistant holds the ranging rod vertically at arm's length so that his body does not obstruct the view of the ranging rod at B.
- The assistant keeps his face towards the surveyor. The surveyor standing at point A directs the assistant to move the ranging rod transverse to the line AB till the three rods at points A, B and C are exactly in a straight line.
- The surveyor then signals the assistant to fix the rod at the point C.
- Then the assistant takes the ranging rod, drags the chain after him and proceeds in the forward direction .
- When the surveyor reaches point C with ranging rod, he directs the assistant to fix the ranging rod at point D at a distance of one chain length from the point C as similar said above. Similar manner, other points are established on line AB.

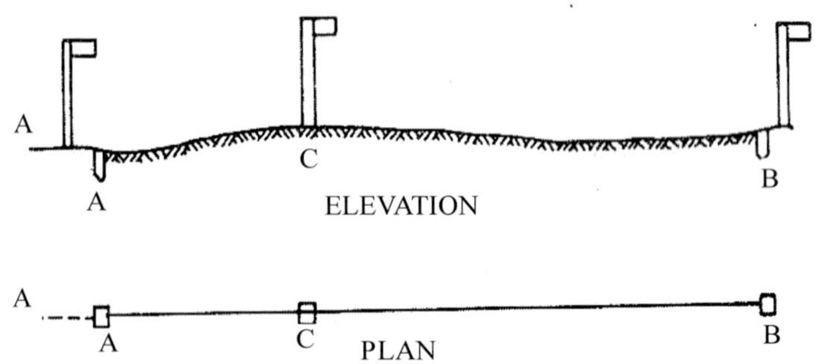

Fig.40.1. Direct ranging a line

Booking field data

- A field book is used for noting the readings when the survey is carried out. The field book is about 22.5 cm long and 12.5 cm wide and opens length wise. Generally single or double red lines are drawn in the middle.
- Chainage lengths are noted in between the two red lines and the space on either side is used for writing the offset lengths and for drawing rough sketches of figures.
- The booking is commenced from the bottom of the page and work upward. The initial and final chainage should be marked with a symbol Ä and capital letters like A, B, C etc., should be used to mark the stations.
- While noting the readings, the surveyor should face the direction of chaining and offsets should be entered on the left or right side as they are situated.
- Each new chain line should be started from a new page. At the commencement of the line in the book is written: the name of the survey

line, the name of station, the symbol perpendicular donating the station, subsidiary station should be indicated by a circle, all around the chain line are entered in the central column

- Offset retain opposite them on the right (or) left of the column according as they are right (or) left of the survey line.

Result:

The length of line AB is ____________

41

Determination of Area of The Field by Triangulation Method of Chain Survey

Aim

- To determine the area of given field by triangulation method of chain survey

Materials required

- Metric chain
- Measuring tape
- Ranging rods
- Chain pins

Principles of triangulation

Triangulation is the basis of trigonometrical or geometric surveys. In chain surveying, the area to be surveyed is suitably divided into a framework consisting of number of triangles (Fig.41.1). The sides of the triangles are directly measured in the field and no angular measurement is taken, it is called triangulation. It can be plotted of the three sides even if the angles are not known. The offsets are taken to locate the details. Plan of the area and details are plotted in the office from the linear measurements taken in the field. Although, the triangulation method is very simple but it is suitable only for small plane areas without much obstruction.

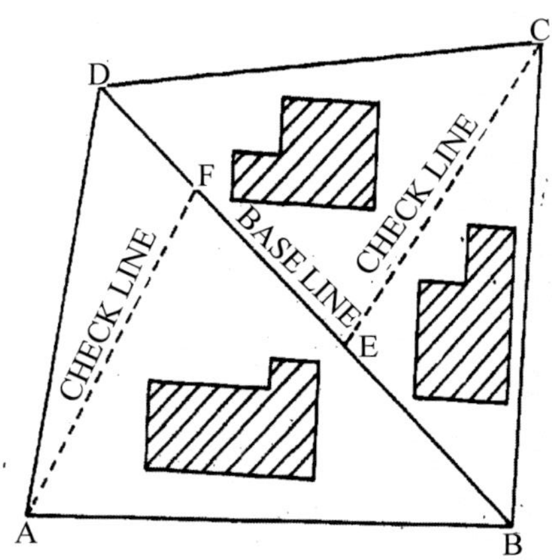

Fig.41.1. Principles chain surveying

Execution of chain survey

Reconnaissance

It is the preliminary inspection of the area to be surveyed to have some idea of the terrain and the principal features of the field. The field should thoroughly examine for suitable positions of the survey stations and survey lines and then decides upon the best possible arrangement of triangles.

Marking survey stations

These are prominent points which are which connect the boundaries of the survey. After completing reconnaissance, the main stations are marked on the ground by fixing a ranging rod at each station.

Running survey lines

The line joining the main survey stations is called survey lines. Survey lines are run to measure the distance between main stations and to locate the adjacent details by offsets.

Measuring base line

It is a long survey line which is run through the middle of the area to be surveyed. The frame work of the triangles is built up on the base line.

Taking offsets

Offsets are lateral distances measured from the survey lines to determine the location of details. These are taken either by perpendicular to chain line or inclined to the chain line.

Check lines

These lines are used to check the accuracy of the frame work and it is measured in the field during surveying. After plotting the plan, the length of the check line is scaled off from the drawing and compared with the measured length in the field.

Tie lines

If the distance of the point of detail from the chain line is very large, long offsets have to be taken. To avoid the long offsets, tie lines are run to locate details

Booking field notes

A field book is used for noting the readings when the survey is carried out.

Field measurement procedure

- The survey stations should be selected in the given shape of field after reconnaissance.
- The ranging rods should be fixed at selected survey stations. Among diagonals of the given shape of the field, the longest is the base line.
- All the four sides AB, BC, CD, DA and base line BD should be measured in the field using chain. The details also should be located by offsets if any in the field.
- All the measurements should be recorded in the field note book. The measurement should be tabulated.
- Chainage lengths are measured along the chain lines for various offsets and points are marked. From these points perpendiculars of suitable lengths are drawn to locate the offsets.
- The intermediate stations are marked on the base line and the frame work of triangles is completed.

Data tabulation

S.No	Survey stations	Length (m)	Details if any	Remarks if any
1	BD			
2	AB			
3	BC			
4	CD			
5	DA			
6	EC			
7	AF			

Plotting procedure

- After the survey is completed, the recorded information should be plotted on a drawing sheet using a suitable scale.
- Depending upon the area covered in the survey and its importance, a suitable scale should be chosen.
- Boundary lines should be drawn leaving a suitable margin all around. While plotting, the base line BD should be drawn first on the sheet.
- The base line which is the mainline in the survey should be suitably located in the map to accommodate the whole plotting easily in the drawing sheet.

- The base line should be most accurately plotted. Draw an arc equal to AD from D and another arc equal to AB from B, and finding out the point of intersection then fix the point A.
- Likewise, the point C is located by the arcs DC and BC. Thus the plan is completed only from the linear measurements.
- To check the accuracy of plotted frame work, check and tie lines should be drawn. The check lines EC and AF (perpendicular lines from the points A and C to the base line BD) should be drawn with the help of setsquare.
- Length of the check lines EC and AF should be scaled-off with scale already chosen. Computes the area of individual triangles and added to get total area.
- The title of the survey, name of the surveyor, date etc., should be written at the right hand bottom corner. The scale is drawn below the map.

Area of the right angle triangle

$$A = \frac{b \times h}{2}$$

Where, b = base, h = altitude

Result

Area of given field by triangulation method of chain survey is _______________

42

Determination of Area of The Given Field by Cross-Staff Survey

Aim

- To find out the area of given field by cross-staff survey method

Materials required

- Cross-staff
- Metric chain
- Measuring tape
- Ranging rods
- Chain pins

Cross-staff

The cross staff survey is a special type of chain survey conducted to locate the boundaries of a field for the purpose of determining the area of the field. Cross-staff is the simplest instrument used for setting out perpendicular i.e taking offsets from a chain line. It is easier and quicker method but not very accurate.

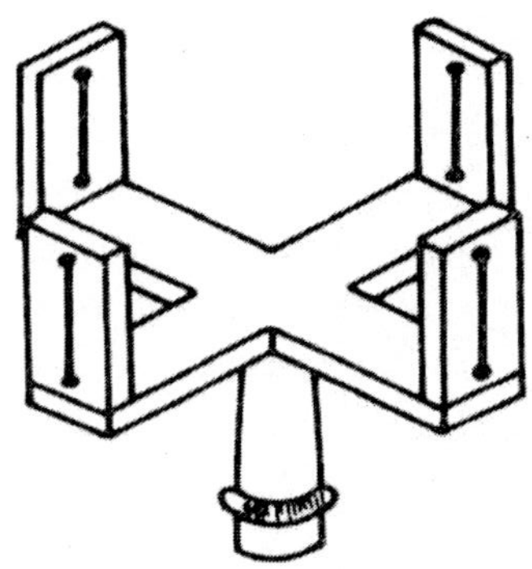

Fig.42.1. Open cross-staff

The simplest form of cross staff is the open cross staff. It consists of two parts (i) the head and (ii) the leg. The head consists of four metal arms with vertical slits. The arms are rigidly fixed in such a manner so that the center of one pair of arms forming a straight line makes right angle with the other pair of arms (Fig.42.1). In one line, one of the slits is narrower than the other. One horse hair is fixed at the center of the wider slit. The object is sighted from the narrow slit in line with the hair. The cross staff is mounted on a wooden staff or pole about 3 cm diameter and 1.5 m long. The pole is provided conical metal shoe so that it can be driven into the ground.

Principles of cross-staff survey

- The area to be measured should be divided into simple polygons such as right-angled triangle and trapezoids because it is easy to find out the area of triangle and trapezoids.
- The base line should be selected so that the offsets (perpendiculars) on its both sides are fairly equal, as far as possible.
- A chain or tape should be used to measure the base line and the perpendicular distances, which are usually quite long.
- A cross staff is used for setting out the perpendiculars on the base line and used for taking offsets.
- To find the foot of the perpendicular from the object the cross staff is held approximately in position and one pair of slits is directed in the direction of the ranging rod fixed at the forward and the chain line.
- The observer then looks through the other pair of slits and sees whether the particular object is bisected or not. If not the cross-staff is moved to and from till the necessary bisection is obtained.
- To set a perpendicular to the chain line at a given point one pair of slits is oriented in the direction of chain line by looking at the ranging rod fixed at the forward and by looking through the other pair of slits ranging rod is fixed in the direction of the line of sight provided by this pair.
- The offsets length should be measured on the left hand side or right hand side of the base line as per position until whole of the area is surveyed.

Procedure for surveying

- ACDBFE is the required closed traverse open field to be surveyed for calculating the area as shown in Fig.42.2.
- From the station A the length of all the opposite corners such as AC, AD, AE, AF and AB should be measured with a chain and the longest distance is considered for laying off the base line.
- In this case AB is the longest and a chain line/ base line running from A to B is laid.
- Offsets to the points C, D, E and F should be laid from the chain line AB by cross-staff and their foot of offsets are G, H, J, K respectively.

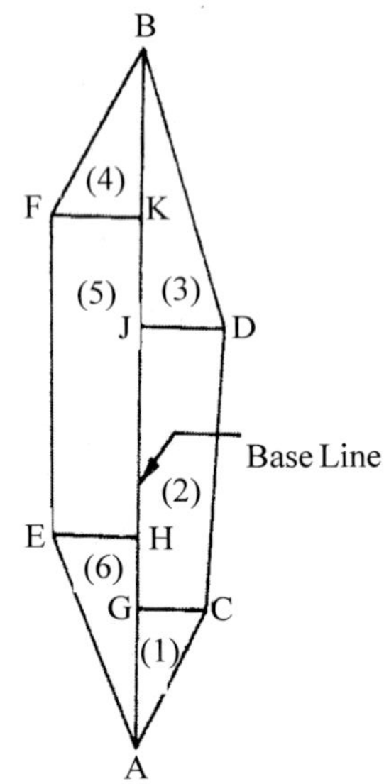

Fig.42.2. Sketch showing cross-staff survey

- All the offset lengths GC, HE, JD and KF should be measured either by chain or tape depending on the length of offsets.
- The distances between all the points AG, GH, HJ, JK and KB should also be measured along the chain line.
- To check the accuracy of the fieldwork, lengths of boundary lines AC, CD, DB, BF, FE and EA should also be measured by direct measurement in the field and compared with the plotted distances. After the field work is over, it should be plotted by taking convenient scale.

Data tabulation

S.No.	Survey stations	Chainage, m				Remarks if any
		Base line, m	Offsets, m	Foot of offsets, m	Boundary lines, m	
1	AB					
2	GC					
3	HE					
4	JD					
5	KF					
6	AG					
7	GH					
8	HJ					
9	JK					
10	KB					
11	AC					
12	CD					
13	DB					
14	BF					
15	FE					
16	EA					

Area calculations: Area of triangles and trapeziums should be calculated and added together to calculate the total area of the given open field.

$$\text{Area of right angle triangle} = \frac{b \times h}{2}$$

$$\text{Area of equilateral triangle} = \frac{\sqrt{3}}{4} \times a^2$$

$$\text{Area of scalene triangle} = \sqrt{s \times (s-a) \times (s-b) \times (s-c)} \quad s = \frac{(a+b+c)}{2}$$

$$\text{Area of trapezium} = \frac{(a+b)}{2} \times h$$

Where, a, b, c = sides; h = altitude (perpendicular distance); s = perimeter

Result

Area of given field by cross-staff survey method is ___________

43

Computation of Area of Irregular Shape of The Field

Aim

- To determine the area of irregular field by use of formulae

Materials Required

- Chain/tape
- Ranging rods
- Chain pins

Theory

Area computation is an important part of surveying. The term 'Area' is meant the projected area on a horizontal plane. The area is computed from the following methods: (i) Graphical method (ii) Instrumental method and (iii) Use of formulae. In graphical method, the area can be computed by dividing the whole area either into triangles or squares. The area of individual triangle is computed from base and the altitude and summation of individual area gives the total area. A tracing paper is divided into number of small squares is placed over the plan. The number of complete squares within the boundary is counted. The squares more than half are counted as full squares and those less than half are neglected. The total number of squares multiplied by the area of each square gives the total area of the field. In instrumental method, the planimeter is used to measure the area of a plotted plan. In use of formulae, computation of area by formulae depends on the shape of the field.

(i) If field with straight boundary

The entire field can be divided into different geometrical figure, such as triangles, rectangles, squares, parallelograms and trapeziums depending upon the shape of the field. Then the area is calculated as follows:

$$\text{Area of recangle, } A = L \times W$$

$$\text{Area of square, } A = a^2$$

$$\text{Area of right angle triangle, } A = \frac{b \times h}{2}$$

$$\text{Area of equilateral triangle, } A = \frac{\sqrt{3}}{4} \times a^2$$

$$\text{Area of scalene triangle, } A = \sqrt{s \times (s-a) \times (s-b) \times (s-c)} \quad s = \frac{a+b+c}{2}$$

$$\text{Area of trapezium, } A = \frac{(a+b)}{2} \times h$$

Where, L = length, W = width, h = altitude, s = perimeter and a, b, c = sides

ii) If field with curved boundary

If sides of the field are bounded by a curved boundary is called irregular field. A base line is drawn through the longest distance in the field and the area is divided into a number of equal parts by perpendicular offset. The ordinate at each of the point of division are drawn and scaled. The following formulae are used for computation of the area of irregular field.

a) Trapezoidal rule

b) Simpson's rule

c) Average ordinate rule

d) Mid ordinate rule

a) Trapezoidal rule

In this method, the area is divided into a series of trapezoids (Fig.43.1). The trapezoidal rule may be stated as follows. The sum of the first and last ordinates plus twice the sum of the intermediate ordinates multiplied by half the common distance between the ordinates gives the required area.

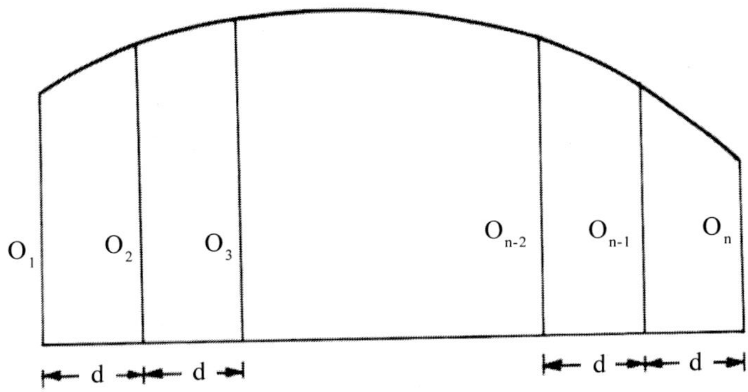

Fig.43.1. Trapezoidal and Simpson rule

$$A = \frac{d}{2}\left[(O_1 + O_n) + 2(O_2 + O_3 + O_4 + ... + O_{n-1})\right]$$

Where,

d = common distance between the ordinates

O_1, O_2, O_n = ordinates

b) Simpson's rule

The rule assumes that the short lengths of the boundaries between the ordinates are parabolic areas and the rule is therefore, called parabolic rule. The rule is stated as follows: The sum of the first and last ordinates is added with four times the sum of all the even ordinates and twice the sum of the remaining odd ordinates. This total sum multiplied by one-third the common distance between the ordinates gives the required area.

$$A = \frac{d}{3}\left[(O_1 + O_n) + 4(O_2 + O_4 + ... + O_{n-2}) + 2(O_3 + O_5 + ... + O_{n-1})\right]$$

Where,

O_3, O_5, O_7, etc = add ordinates

O_2, O_4, O_6, etc = even ordinates

d = common distance between the ordinates

c) Average ordinate rule

The average of the all ordinates multiplied by the length of the base line gives the required area (Fig.43.2).

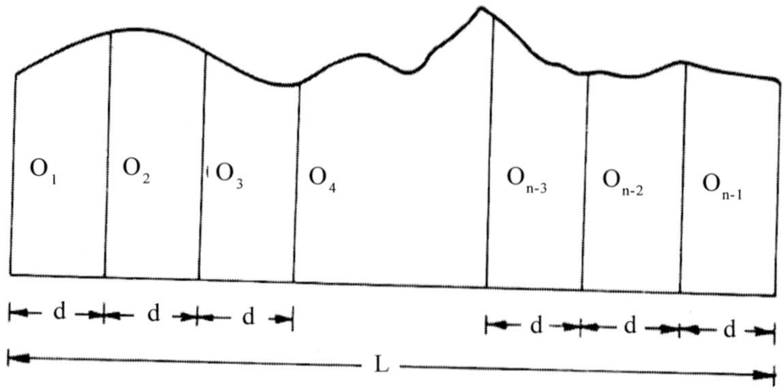

Fig.43.2. Average ordinate rule

$$A = \frac{L}{N+1}\left(O_1 + O_2 + O_3 + O_4 \ldots\ldots\ldots\ldots + O_n\right)$$

Where,

$O_{1,}$ $O_{2,}$ $O_{3,}$ etc = ordinates

L = length of the base line

N = number equal parts

d) Mid-ordinate rule

In this method, the whole area is divided into number of equal divisions (Fig.43.3). The mid-ordinates are measured at the mid points of each division and then area is calculated by the formula:

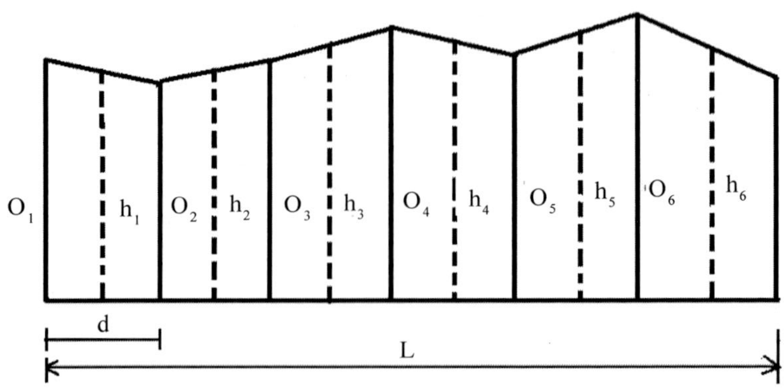

Fig.43.3. Mid ordinate rule

$$h_1 = \frac{O_1 + O_2}{2}$$

$$h_2 = \frac{O_2 + O_3}{2}$$

$$A = L \times \left(\frac{h_1 + h_2 + h_3 + h_4 + \ldots\ldots + h_n}{n} \right)$$

(OR)

$$A = d \times (h_1 + h_2 + h_3 + h_4 + h_5 + h_6)$$

Where,

L = length of the base line

d = common distance between the ordinates

n = number of equal parts into which the base line is divided at a distance of 'd'

$O_1, O_2, O_3, \ldots$ = ordinates

$h_1, h_2, h_3, \ldots$ = ordinate at the midpoint of each division

Worked example

The following perpendicular offsets 6.4, 7.2, 8.5, 9.6, 12.5, 11.4, 8.8 m were taken at 10 m intervals from a base line to a field boundary. Calculate the area enclosed between the baseline, the field boundary and the first and last offsets by (i) Trapezoidal rule (ii) Simpson's rule (iii) Average ordinate rule and (iv) Mid-ordinate rule and

Trapezoidal rule

$$A = \frac{d}{2}\left[(O_1 + O_n) + 2(O_2 + O_3 + O_4 + \ldots\ldots + O_{n-1})\right]$$

$$A = \frac{10}{2}\left(6.4 + 8.8 + 2(7.2 + 8.5 + 9.6 + 12.5 + 11.4) = 568 m^2\right)$$

Simpson's rule

$$A = \frac{d}{3}\left[(O_1 + O_n) + 4(O_2 + O_4 + \ldots + O_{n-2}) + 2(O_3 + O_5 + \ldots + O_{n-1})\right]$$

$$A = \frac{10}{3}\left[(6.4 + 8.8) + 4(7.2 + 9.6 + 11.4) + 2(8.5 + 12.5)\right] = 566.7 m^2$$

Average ordinate rule

$$A = \frac{L}{N+1}\left(O_1 + O_2 + O_3 + O_4 .... + O_n\right)$$

$$A = \frac{60}{6+1}\left(6.4 + 7.2 + 8.5 + 9.6 + 12.5 + 11.4 + 8.8\right) = 552m^2$$

Mid-ordinate rule

$$A = L \times \left(\frac{h_1 + h_2 + h_3 + h_4 + ... + h_n}{n}\right)$$

$$h_1 = \frac{O_1 + O_2}{2} = \frac{6.4 + 7.2}{2} = 6.8$$

Similarly,

$h_2 = 7.85$; $h_3 = 9.05$; $h_4 = 11.05$; $h_5 = 11.95$; $h_6 = 10.10$

$$A = 60 \times \left(\frac{6.8 + 7.85 + 9.05 + 11.05 + 11.95 + 10.10}{6}\right) = 568m^2$$

Exercise

Problem 1: The perpendicular offsets such 0, 3.5, 5, 7.6, 10.8, 15, 20, 16.7, 12.5, 8.8, 4.5, 0 m were taken at 10 meter intervals from a base line to a field boundary. Calculate the area enclosed between the baseline, the field boundary and the first and last offsets by (i) Trapezoidal (ii) Simpson's (iii) Mid-ordinate and (iv) Average ordinate rules.

Problem 2: The following perpendicular offsets were taken at 15 meter intervals from a base line to a field boundary. 5, 7.6, 10, 15.7, 20.5, 30, 26.6, 22, 17.4, 12.5, 8, 3.5, m Calculate the area enclosed between the baseline, the field boundary and the first and last offsets by (i) Trapezoidal (ii) Simpson's (iii) Mid-ordinate and (iv) Average ordinate rules.

44

Studies on Prismatic Compass and Angular Measurement

Aim

- To identify different parts of prismatic compass and to know their functions.
- To practice of taking the angular measurement.

Compass survey principle

Compass surveying is a branch of surveying in which directions of survey lines are determined with a compass and the lengths of the lines are measured with a chain or tape. In practice, the compass is generally used to run a traverse. In compass surveying, a traverse consists of a series of straight lines connected together to form an open or closed polygon as shown in Fig.44.1. The points A, B, C, D and E defining the ends of the traverse lines are called traverse station or traverse points. The lengths of the traverse lines are measured with a chain or tape. At each traverse station, where the traverse lines meet, the angle is measured with a compass.

Prismatic compass

The prismatic compass is a magnetic compass. It is used for the determination of whole circle bearings of the lines. The prismatic compass consists of a cylindrical metal or brass box of about 80 to 120 mm diameter. In the center there is a pivot, which carries a magnetic needle. The magnetic needle is attached to a graduated aluminium ring with the help of an agate cap. The box is fitted with a glass disc at its top. When the compass is not in use, the box is covered with a brass lid.

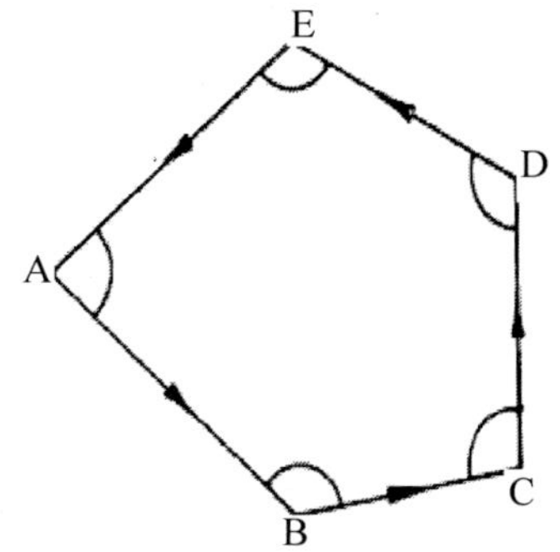

Fig.44.1. Traverse anticlockwise

The aluminium ring is graduated from 0° to 360° with a minimum of half-degree division. Although reading is taken in the clockwise direction, the divisions are made in the anticlockwise direction with 0° at the magnetic south and 180° in the magnetic north. That means 90° and 270° divisions are in the west and east respectively. The graduations are engraved inverted on the aluminium ring. The readings are taken through the reflecting prism placed exactly opposite the sight vane. The prism is kept at 45° with its horizontal and vertical faces slightly convex so as to magnify the image of graduations. Thus the prism reflects the magnified image of the graduations vertically below. Prism cap protects the prism from dust, moisture etc. There is a sighting vane or slit through which the object is seen through the object vane which carries a horse hair (Fig.44.2).

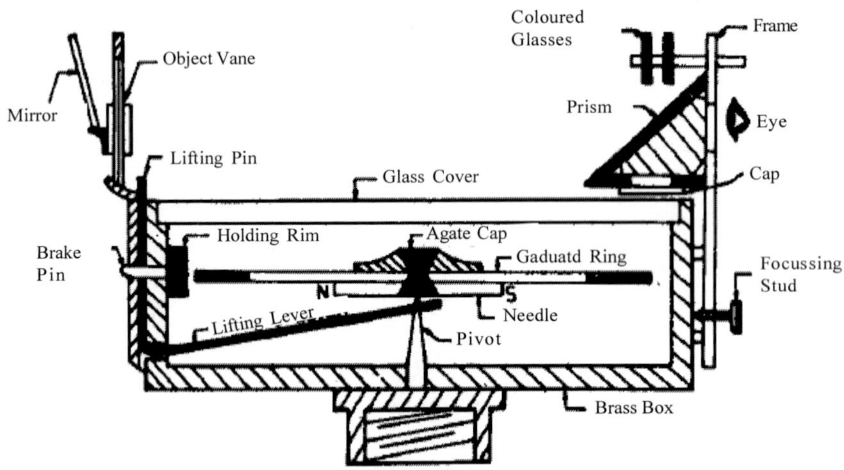

Fig.44.2. Prismatic compass

The prism can be adjusted to the eyesight by the focusing screw. When not in use the object vane is folded on the glass cover, which presses against a lifting pin. Then the lifting lever lifts the magnetic needle off the pivot, which is saved from undue wear. To bring the ring quickly to rest, brake pin is pressed which presses a spring brake to come in contact with the ring. When very high or low objects are to be sighted, the sliding reflecting mirror can be used. For working in very bright sun or sighting luminous object, colored glasses should be placed between the slit and the object vane.

Measurement of bearing

i) Leveling the compass

The prismatic compass is mounted on a tripod by screwing it to a vertical spindle in a ball and socket joint. With this arrangement, the compass can easily be leveled, rotated in a horizontal plane and fixed in any position.

ii) Centering the compass

The compass is placed approximately over the station and a stone is dropped from below the center of the compass. For accurate centering, the stone should fall over a peg fixed at the station. By means of the ball and socket arrangement, the compass is leveled so that the graduated ring along with the magnetic needle can swing freely.

iii) Observing the bearing

After centering and leveling, the prism is adjusted till the graduations on the ring are clearly visible when looked through the prism. The compass box is turned until the ranging rod to be sighted is bisected by the horse hair when looked through the slit above the prism. As soon as the ring comes to rest, reading is taken through the prism sighting ranging rod.

iv) Meridian

The direction of a line is defined by the horizontal angle, which the line makes with a reference line. Generally, the directions are measured clock wise from the reference line. The fixed line of reference is called a meridian.

v) True bearing

The bearing of a line is the horizontal angle, which it makes with a reference line (meridian). The bearings indicate the directions with respect to the meridian (Fig.44.3a). The true bearing of a line is the horizontal angle between the true meridian and the line. The true bearing of a line is also called as the azimuth.

vi) Magnetic bearing

The magnetic bearing of a line is the horizontal angle, which the line makes with the magnetic north (Fig.44.3b). These are determined with a prismatic compass as the whole circle bearing.

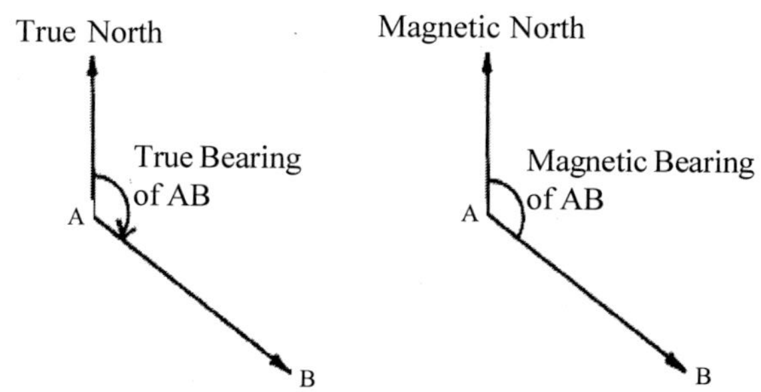

Fig.44.3. True and magnetic bearing

vii) Whole circle bearing

The whole circle bearing of a line is the horizontal angle, which the line makes with the north end of the reference meridian. The bearing of a line is measured clockwise from the north end of the reference meridian. The prismatic compass is used for measuring the whole circle bearing. The WCB of a line may vary from 0° to 360°. In Fig.44.4, the whole circle bearing of the lines OA, OB, OC and OD are respectively 45°, 150°, 240° and 330°.

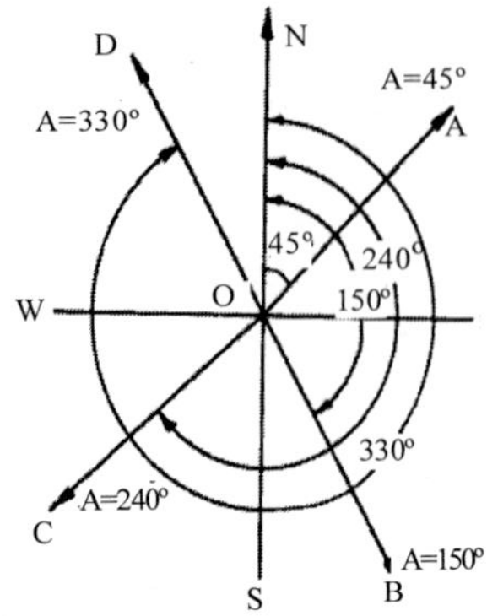

Fig.44.4. Whole circle bearing system

viii) Fore bearing

Any straight line has two diametrically opposite directions, depending upon the direction of arrow (Fig. 44.5a). The bearing of a line taken in the direction of the progress of survey work is known as fore bearing (FB).

(x) Back bearing

The bearing of line taken in the direction opposite to the direction of the progress of survey work is known as back bearing (BB). To avoid confusion, the arrows should be marked to indicate the direction of the progress of survey (Fig.44.5b).

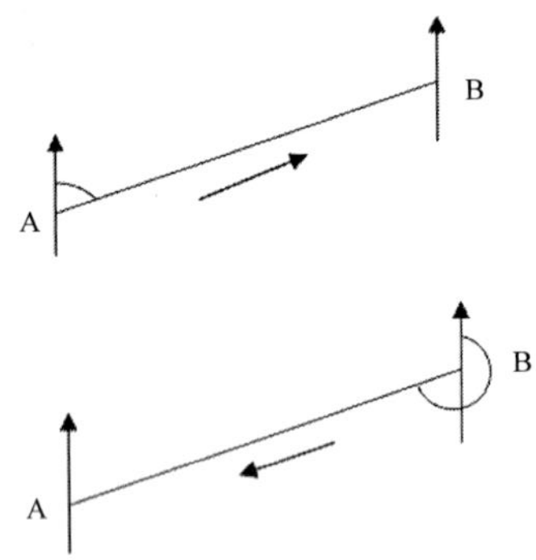

Fig.44.5. Fore bearing and back bearing

Determination of back bearing from fore bearing

The back bearing of a line may be determined if its fore bearing is given and vice versa, remembering that they always differ by 180°.
The following rule is applied for conversion of back bearing and fore bearing. If the fore bearing of a line is given as the whole circle bearing:

Back bearing = Fore bearing + 180°, if F.B < 180°

Back bearing = Fore bearing - 180°, if F.B > 180°

Observation

i) Fore bearing of line AB is 75°. Find out the back bearing.

Back bearing = 75° + 180° = 255°

ii) What would be back bearing line BA if fore bearing of line AB is 240°?

Back bearing = 240 - 180 = 60°

iii) Back bearing of line BA is 320°. What is the value of fore bearing?

Back bearing = Fore bearing + 180°

Fore bearing = 320 - 180 = 140°

45

Determination of Area of The Open Field by Radiation Method of Compass Survey

Aim

- To determine the area of the given field by radiation method using a prismatic compass.

Materials required

- Prismatic Compass
- Tripod
- Measuring tape
- Arrows
- Ranging Rods

Field work procedure

After conducting the reconnaissance survey, the surveyor should fix up the positions of all the station around the given field. The approximate centre of the field (say 'O') should be located and a chain pin should be fixed on this point.

Centering

- The tripod should be laid at this point (O). By moving the legs of the tripod suitably, centering should be done.
- Centering should be checked by dropping a stone so that it falls on the top of the peg.
- The prismatic compass should be mounted on the tripod by screwing it to a vertical spindle in a ball and socket joint.

Leveling

- After centering of the compass, raise the sighting vane and prism of compass. By means of the ball and socket arrangement, the compass can easily be levelled, rotated in a horizontal plane and fixed in any position so that the graduated ring along with the magnetic needle can swing freely.

Focusing

- After centering and leveling, the prism should be adjusted till the graduations on the ring are clearly visible when looked through the prism.
- The compass box should be turned until the ranging rod to be sighted is bisected by the horse hair when looked through the slit above the prism.
- As soon as the ring comes to rest, take the reading through the prism sighting the ranging rod simultaneously.

Observing bearings

- The compass should be rotated until the line of sight is bisects the object at 'A' (the object vane i.e ranging rod at station 'A' and slit of the sighting vane are in the same line).
- Then read the graduated ring through prism accurately and note down the bearing of line OA. The reading directly gives the magnetic bearing of 'OA' in whole circle bearing system (Fig.45.1).
- Follow the same procedure to observe the magnetic bearing of lines OB, OC, OD and OE by rotating the compass box in the respective directions.
- The readings indicates that the angles with the line, OA, OB, OC, OD and OE makes with the north line.
- Then with the help of chain, measure the distances OA, OB, OC, OD and OE on ground, after proper ranging.

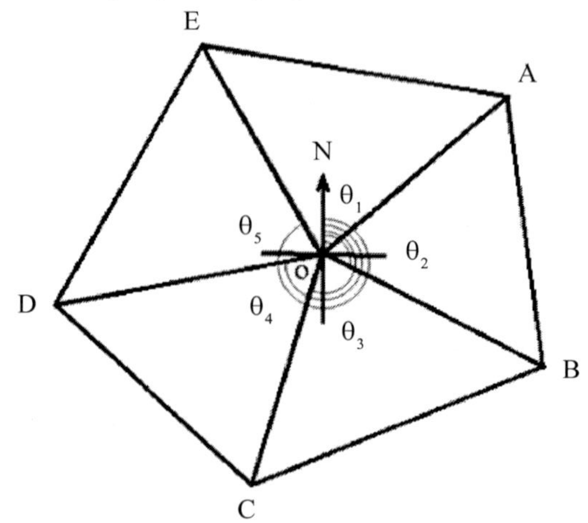

Fig.45.1. Sketch showing radiation method

Data tabulation

Sl.No	Survey stations	Fore Bearing	Distance (m)	Remarks if any
1	OA	Q_1		
2	OB	Q_2		
3	OC	Q_3		
4	OD	Q_4		
5	OE	Q_5		

Plotting procedure

- For plotting the field measurement, convenient scale should be selected depending on the length of survey lines from the centre of the field.
- By knowing the observed bearings, the direction of survey lines OA, OB, OC, OD and OD should be plotted from the centre point 'O'.
- Draw an arc on the line OA from the centre point 'O' corresponding to measured length of line OA and find out the point of intersection then fix the point A.
- Follow the similar procedure to find out the point of intersection on the lines OB, OC, OD and OD.
- Join the station points A, B, C, D, E which will indicate the plan of the plot. The area enclosed by the plot may be calculated by splitting the plan into number of triangles such as OAB, OBC, OCD, ODE and OEA.
- The length of sides of the triangles may not be equal thus use scalene triangle formula to determine the area.

Area calculations

The area can be calculated by following formula:

$$A = \sqrt{s \times (s-a) \times (s-b) \times (s-c)}$$

$$S = \frac{a+b+c}{2}$$

Where,

s = perimeter

a, b, c = sides

Total area = $A_1 + A_2 + A_3 + A_4 + A_5$

Result

Area of the given field by radiation method using a prismatic compass is __________.

46

Determination of Area of Given Field and Included Angle of the Closed Polygon by Compass Survey

Aim

- To determine the area of the field by intersection method using a prismatic compass.
- To determine the included angle of the closed polygon.

Materials required

- Prismatic Compass
- Tripod
- Measuring tape
- Arrows
- Ranging Rods

Field procedure

- The area to be surveyed should be a closed polygon. The ranging rods should be fixed at different corners of the polygon i.e. A, B, C, D and E such that they should be mutually visible and may be measured easily (Fig.46.1).
- The prismatic compass should be set on the tripod stand at point 'A'. The compass should be properly cantered & levelled over the staring station A.
- Ensure that the north-south direction of magnetic needle and the compass should be turned towards the ranging rod at E till the rod is ranged through sighting slits and objective vane attached with horse hair and reading on prismatic compass should be noted down. It is back bearing of the line AE.
- After taking back bearing, turn the compass towards the ranging rod at B till bisects the horse hair at objective vane and ranging rod where reading should be noted down on compass. It is fore bearing of line AB.

- The traverse line AB should be ranged as in chain surveying, and its length is measured with a chain or a tape.
- Then the compass along with tripod should be shifted to the station B where the compass should be again cantered and levelled.
- The fore bearing of the line BC and back bearing of the line AB are measured as stated in previous. The traverse line BC should be ranged and chained, and offsets are taken if any.
- Similarly, the compass should be shifted to the station C, D and E, respectively, and the processes of taking bearings, ranging, chaining and offsetting are repeated. Thus the traverse is completed.
- So, fore bearing and back bearing of each line is measured at all the stations. The fore bearing and back bearing of each line should differ by 180°.
- If the difference between the fore bearing and the back bearing of the line is not 180°, there is an error.
- If the error is more than permissible limit, the bearings of that line should be measured again. If the error is within the limit, it can be adjusted by Bowditch's rule.
- In a closed traverse, the theoretical sum of the included angles should be equal to $(2N-4)\times 90°$ Where, N is the number of sides.

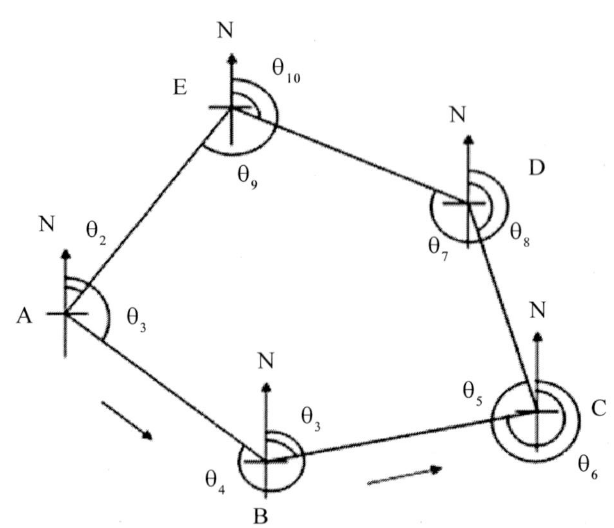

Fig.46.1. Sketch showing intersection method of compass survey

Data tabulation

S.No	Survey line	Length (m)	Fore bearing (FB)	Back bearing (BB)	Remarks
1.	AB				
2.	BC				
3.	CD				
4.	DE				
5.	EA				

Plotting procedure

- For plotting the field measurement, convenient scale should be selected depending on the length of traverse lines. Draw a line indicating the magnetic meridian (reference line) at 'A'.
- By knowing the observed fore bearing and back bearing, direction of the survey lines AB and AE should be plotted from the point A.
- Draw an arc on the line AB and AE from the point A corresponding to measured length of line AB and AE respectively and find out the point of intersection then fix the point B and E.
- Then draw a line indicating the magnetic meridian (reference line) at 'B'. By knowing the observed fore bearing and back bearing, direction of the survey lines BC should be plotted from the point B.
- Draw an arc on the line BC from the point B corresponding to measured length of line BC and find out the point of intersection then fix the point C.
- Similarly, other points C, D, E should be located and draw all the remaining lines and complete the closed traverse.
- Check the accuracy by comparing the difference between fore bearing and back bearings taken from first and last stations. If they differ by 180°, the plotting is correct.
- If they not differ by 180°, there may be a closing error which should be adjusted graphically (Bowditch's Rule).

Calculation off included angle

When two lines meet at a point two angles namely interior angle and exterior angle are formed. The sum of these angles is equal to 360°. The following rules may be followed to calculate included angles between two lines whose bearings are given:

Case: I

If the whole circle bearings of the two lines are measured from a common point, the included angle between the lines equal to the difference in the two bearings, if the value is less than 180°. Otherwise the interior angle is obtained by subtracting the difference from 360°.

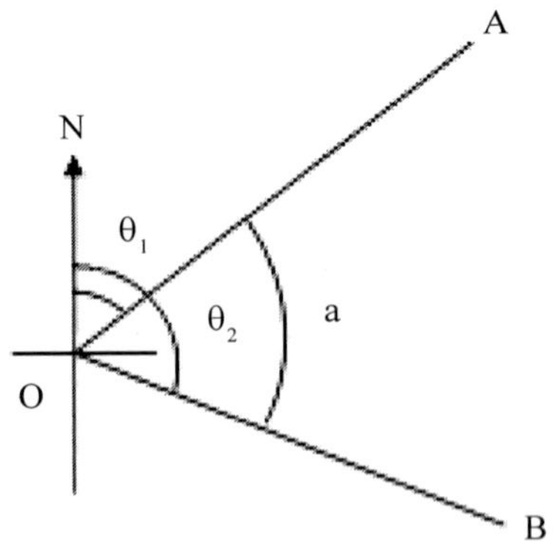

Included angle, $\alpha = \theta_2 - \theta_1$

Case: II

When the bearings of the two lines are not measured from their common point, first determine their fore bearings at their common point if their back bearings are measured at different point by applying the rules of the back bearing.

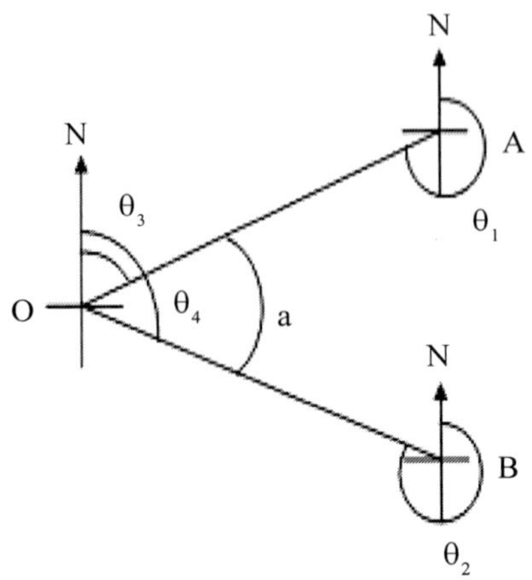

θ_1 = BB of the line OA, then its FB, $\theta_3 = \theta_1 - 180°$
θ_2 = BB of the line OB, then its FB, $\theta_4 = \theta_2 - 180°$
Thus, Included angle, $\alpha = \theta_4 - \theta_3$

Angle A, Bearing of AB – bearing of AE

If the answer is greater than 180º, it is exterior angle.

(To get interior angle, subtract it from 360º)

Angle B, Bearing of BC – Bearing of BA

Angle C, Bearing of CD – Bearing of CB

Angle D, Bearing of DE – Bearing of DC

Angle E, Bearing of EA – Bearing of ED

Included angle = Angle A + Angle B+ Angle C+ Angle D+ Angle E

Area calculations

- To calculate the area of the closed polygon, the plot should be divided into number of triangles. Therefore, draw a line AD and BD to get triangles BCD, ABD and ADE.
- Area should be calculated by using, corresponding formula and summarise all the areas to get total area of the given field.

The area should be calculated by following formula:

$$A = \sqrt{s \times (s - a \times) \times (s - b) \times (s - c)}$$

$$S = \frac{a + b + c}{2}$$

Where,

s = perimeter

a, b, c = sides

Total area = Area of triangle ADE + Area of triangle ABD + Area of triangle BCD

Result

The included angle of the closed polygon is ______________.

Total area of the given field by intersection method using a prismatic compass is __________.

47

Determination of Area of The Field by Radiation Method of Plane Table Survey

Aim

- Studies on plane table survey instruments
- To determine the area of given open field by radiation method of plane table survey.

Plane table survey

Plane table surveying is a simple method of surveying for determination of area of field. The plane table survey is quite suitable for plotting small-scale and medium scale mappings directly in the field where great accuracy is not required. It is also used for plotting the topographical maps in the field. The plane table survey is generally more rapid and less costly than other types of survey.

In this survey, a plane table is used for taking the measurements and for plotting the plan in the field. Plane table surveying is a method in which the field observations and plotting of the plan proceed simultaneously. The plotted map can be compared with the actual features in the field. Errors in measurements and plotting can be easily detected in the field by taking suitable check lines.

Components of plane table survey

- Plane table and tripod
- Alidades
- Plumbing fork
- Plump bob
- Trough compass
- Level tube
- Drawing sheet

i) Plane table with tripod

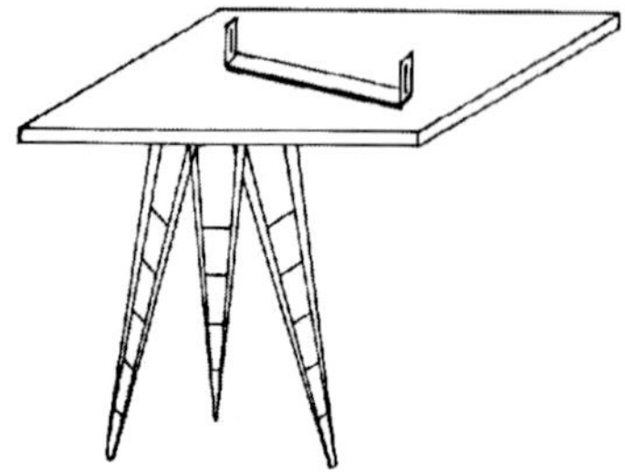

Fig.47.1. Simple plane table with tripod

A plane table is used for taking the measurements and for plotting the plan in the field. It consists of a drawing board made of well-seasoned wood. The board is mounted on the tripod when required for plane table surveying. The size is generally 75 cm × 60 cm and thickness is about 20 mm. An aluminium racer ring is provided on the underside at the center of the board. There is a threaded socket in the racer ring, which receives a screw passing through the head of the aluminium casting fitted to the tripod (Fig.47.1).

A screw is provided with a wing nut at its lower end. The table can be set in any direction in the horizontal plane by means of the wing nut. When the wing nut is tightened, the plane table is clamped firmly to the tripod and it can be rotated in the horizontal plane. After mounting the board on the tripod, the plane table is leveled by moving and adjusting the tripod legs. Eye–judgment is used for checking the leveling. The height of the tripod is about 1.2 m so that the surveyor can easily bend over the table while taking measurements and plotting the plan.

ii) Plumbing fork

A plumbing fork is a U-shaped piece of metal. It is used for the centering of the plane table over the station and transferring of the ground point on to the drawing sheet The end of the upper arm of the fork is pointed, and it is placed on the drawing paper fixed to the plane table (Fig.47.2). A plumb bob is attached to the end of the lower arm of the fork. The point of the upper arm and the plumb line are in the same vertical line.

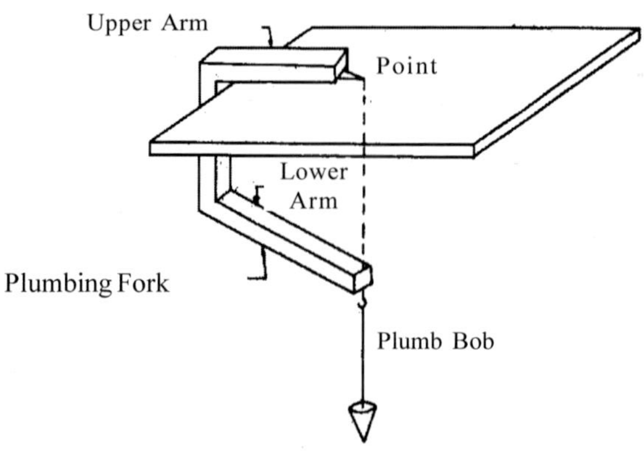

Fig.47.2. Plumbing fork

iii) Plumb bob

The plumb bob is always freely suspended towards the gravity and it is indicates the direction of the vertical line. A plumb bob is a shape of conical ball made of bronze and has a fine, hardened steel point. There is a hook at the top for attaching a string of nylon (Fig.47.3).

The plumb bob is used for centering the table. A plumb bob is attached to the end of the lower arm of the fork. When the point of the plumbing fork is kept over the plotted point and the plumb bob is exactly over the station peg, the plane table is centered over the station.

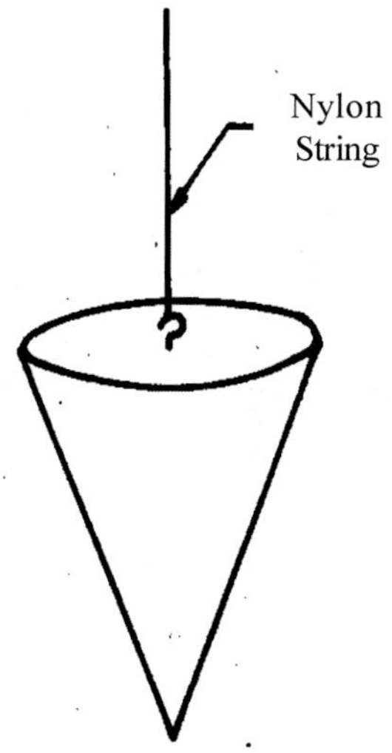

Fig. 47.3. Plumb bob

iv) Alidade

The alidade is used for sighting the objects and drawing the lines. It is a straight edge ruler about 45 cm long, made of a metal or wood. One of the edges is beveled and graduated. The beveled edge of the ruler is also known as the fiducial edge (Fig.47.4). The alidade is provided with a sight vane at each end. One of the sight vanes is provided with a narrow slit and is used as an eye vane and other sight vane is open and it carries a thin wire at its center. The vanes have hinges at the lower end so that they can be folded down on the ruler when not in use. The line of sight of the alidade is in the same as that of the fiducial edge or in a plane parallel to it.

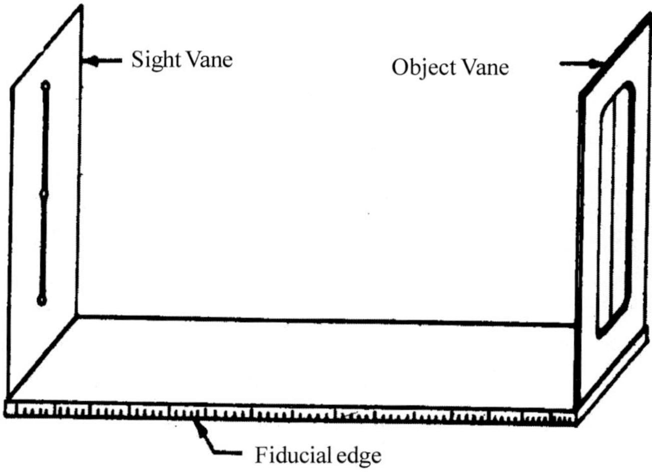

Fig.47.4. Alidade

v) Trough compass

A trough compass is used to orient the plane table with respect to the magnetic meridian. It is used for marking the north line on the drawing sheet of the plane table. It is a long, narrow rectangular box and covered with glass. Inside the box, at its center, there is a magnetic needle resting on the pivot (Fig.47.5). At the extremities of the trough compass, there are graduated scales with zero ay the center and marking upto 5° on either side of the zero line. When the compass is so kept that the two ends of the magnetic needle point to 0° - 0° of the graduated scale, a line drawn parallel to the edge of the trough compass is along the magnetic meridian.

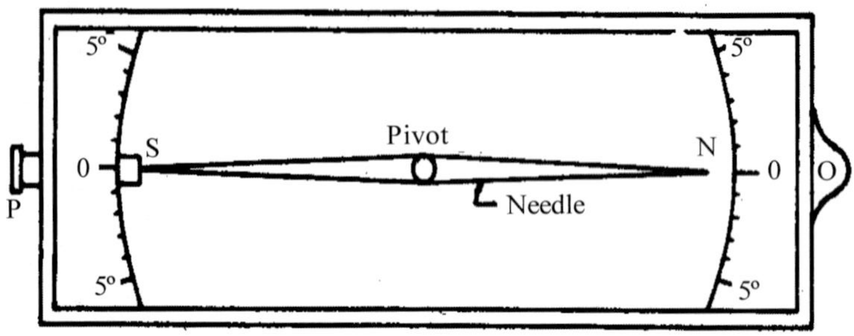

Fig.47.5. Trough compass

vi) Level tube

A level tube is used to check the plane table is properly leveled. The base of the level tube is flat so that it can be laid on the table. Leveling is checked by placing the level tube on the board in two positions at right angles to each other. When the bubble of the level tube is in the center, the table is leveled.

vii) Drawing sheet

The drawing sheet is fixed on the top of the board. The paper is fixed on the plane table by means of clips or suitable clamps or with an adhesive tape. It should be of good quality and able to withstand the rubbing of the alidade. For accurate and permanent record, fiberglass sheet or paper backed with aluminium sheet are used.

Procedure for setting up the plane table

Reconnaissance of the area should be done before selecting the plane table stations. Obstructed sights can be avoided by selecting the plane table stations. The plane table stations should be on the firm, level ground. The plane table is set up after selecting the plane table station.

i) Fixing the plane table

The legs of the tripod should be spread, and the shoes of the legs should be pushed into the ground smoothly. The top of the tripod should be at a convenient height, about 1.2 m, depending upon the height of surveyor, so that he can conveniently work on it. The board is fixed to the tripod and the clamp is tightened. By moving the legs of the tripod, the plane table is leveled approximately.

ii) Leveling the plane table

A level tube is used to check the plane table is properly leveled. The base of the level tube is flat so that it can be laid on the table. Leveling is checked by placing the level tube on the board in two positions at right angles to each other. When the bubble of the level tube is in the center, the table is leveled.

iii) Centering the plane table

Centering is the process of setting up the plane table such that the plotted point 'O' on the drawing sheet corresponding to the ground station 'O' is exactly over the station. A plumbing fork is used for checking the centering. The pointed end of the upper leg of the plumbing fork is made to coincide with the plotted point 'O' and the table is shifted physically until the plumb bob is exactly over the ground station (Fig.47.6).

iv) Orientating the plane table

This is the process of aligning the plane table by rotating it in the horizontal plane such that all the plotted lines are parallel to the corresponding lines on the ground. Orientation of the plane table can be done using a trough compass, by back sighting.

v) Back sight

It is a sight taken from a plane table station to another station whose position had already been plotted on the drawing paper. For taking a back sight to station (say) B when the plane table is centered over a station A, the alidade is placed along the plotted line AB. the plane table is rotated until the station B is bisected. Thus the plane table is oriented by back sighting.

vi) Foresight

It is a sight taken from a plane table station to another station whose position had not already been plotted. It is taken to locate the position of the forward station.

vii) Observation of reading

- The ranging rods should be erected at selected survey stations A, B, C, D and E. The plane table should be set over the station 'O' which is centre of the field.
- Then it should be leveled and centered over the ground station at 'O' and a drawing sheet should be fixed on the table.
- The north line (N) is marked on the right-hand top corner of the sheet with trough compass. The alidades is kept on the drawing sheet, with the zero of the fiducial edge set on the point 'O'. This is called pivoting of the alidade.
- With the alidade pivoted on the point O, the alidade is rotated so that the line of sight passes through the object A as shown in Fig.47.7.
- A line should be drawn on the paper along the fiducial edge of the alidade. The line represents the direction from the station O to A.
- The distance OA should be measured in the field with a tape or a chain. The measured distance should be then plotted to scale as 'oa' along the line already drawn. Thus the point 'a' is located on the map.
- Repeat the above procedure for the location of points B, C, D and E. Join the plotted points a, b, c, d and e as shown in Fig.47.7.
- It is conventional to use capital letters O, A, B, etc. for the ground stations and the lower case letters o, a, b, etc. for the plotted points corresponding to the ground stations.

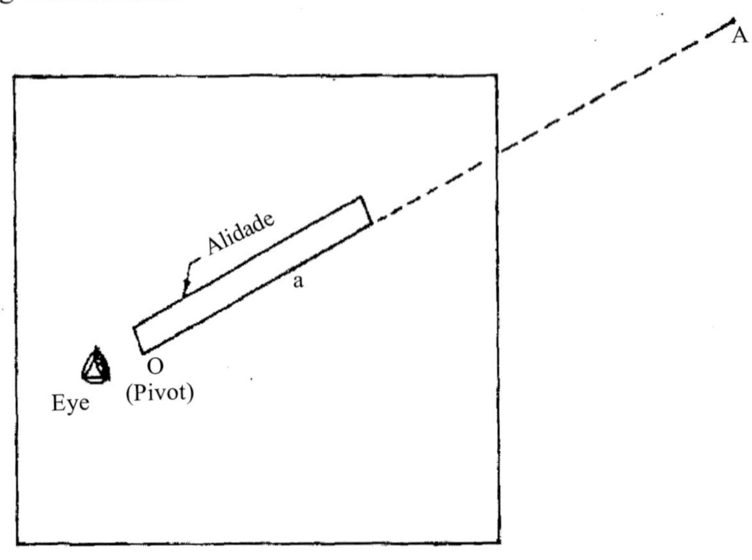

Fig.47.6. Sketch showing locating survey station

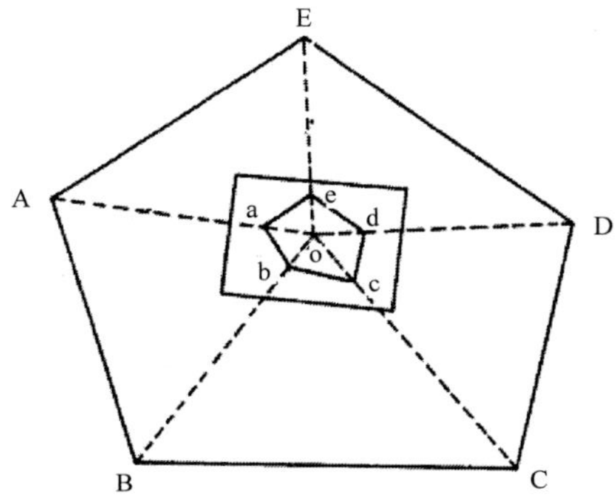

Fig.47.7. Sketch illustrating plane table survey

Data tabulation

S.No	Survey lines	Distance (m)	Remarks, if any
1	OA		
2	OB		
3	OC		
4	OD		
5	OE		

Calculation of area

The area of triangles OAB, OBC, OCD, ODE and OEA should be calculated by corresponding formula and summarise all the areas to get total area of the given field.

The area can be calculated by following formula:

$$A = \sqrt{s \times (s-a) \times (s-b) \times (s-c)}$$

$$S = \frac{a+b+c}{2}$$

Where,

s = perimeter

a, b, c = sides

Total area = Area of OAB + Area of OBC + Area of OCD + Area of ODE + Area of OEA

Result

Total area of the given field by radiation method of plane table survey is ________.

48

Determination of Area of The Field by Intersection Method of Plane Table Survey

Aim

- To determine the area of given field by intersection method of plane table survey.

Materials required

- Tripod
- Plane table
- Plumbing fork
- Level tube
- Magnetic compass
- Alidade
- Measuring tape
- Ranging rods
- Drawing sheet
- Pins, pencil and eraser

Theory

The method of intersection is commonly used for location of details. The method is ideally suited for locating the inaccessible points when the linear measurements are not possible. The method is also used for locating points which are to be subsequently used as the plane table station. The method is generally used when it is difficult or impossible to measure the horizontal distances, as in hilly area. In this method, the positions of the points are fixed on the sheet by the intersection of the rays drawn from two instrument stations. The line joining these stations is termed as base line which is only linear measurement taken in this method.

Procedure

- Select two points A and B such that all points to be plotted are visible from both points as shown in Fig.48.1.
- The line joining the points A and B is called base line, which is only linear measurement taken in this method.
- Measure the distance between A and B, and plot with suitable scale.
- Set up the plane table at A and level it and mark a suitable point 'a' on the drawing paper to represent the ground station A, using the plumbing fork.
- Mark the direction of the magnetic meridian on the top corner of the drawing sheet by means of a trough compass.
- Pivot the alidade on the point 'a' and sight the station B, and draw a ray in that direction.
- Measure the distance AB and scale off the distance 'ab' to locate the point 'b'. Thus the base line ab is plotted.
- Pivot the alidade on 'a', sight the stations P, Q, R, S to be plotted and draw the rays along the fiducial edge of the alidade towards them. Mark the respective lines by letters p', q', r', s'.
- Shift the instrument to station B, set it over the B, and level and exactly centre the table using on B using the plumbing fork.
- Orient the table by placing the alidade along b, turn the table until the line of sight strikes A, and then clamp it.

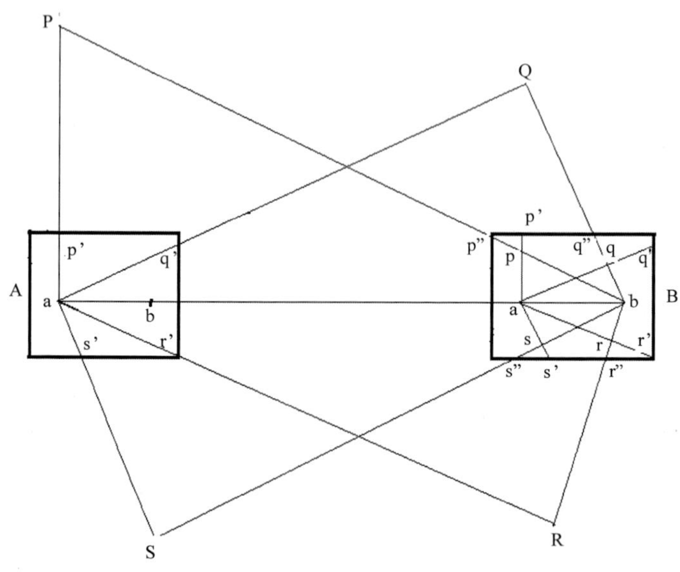

Fig.48.1. Intersection method of plane table survey

- Pivot the alidade on b, and sight the stations P, Q, R, S and draw the rays along the stations, and mark the respective lines by p”, q”, r”, s”.
- The intersections of the ray ap’, aq’, ar’, as’ drawn from ‘a’ and the rays bp”, bq”, br”, bs” drawn from ‘b’ gives the position of the points p, q, r, s.
- Join p, q, r, s as shown in figure and find area.

Area calculation

- Measure distance of lines ap, pq,qb, br, rs, sa and scale up.
- Calculate the area using suitable triangle formulae.

Result

Area of given field by intersection method of plane table survey is _________

49

Determination of Area of The Field by Traversing Method of Plane Table Survey

Aim

- To determine the area of given field by intersection method of plane table survey.

Materials required

- Tripod
- Plane table
- Plumbing fork
- Level tube
- Magnetic compass
- Alidade
- Measuring tape
- Ranging rods
- Drawing sheet
- Pins, pencil and eraser

Theory

A traverse is series of straight lines connected together. In plane table traverse, the angles are directly plotted without measuring them. It is generally used for locating topographical details directly on the plane table. As in a compass traverse, the traverse stations are first selected. The plane table is set successively on all the traverse stations and a back sight is taken on the preceding station and a fore sight is taken on the following station. In a plane table traverse, the measured lines are plotted directly on the drawing paper to a suitable scale in the field.

Procedure

- Select the traverse stations A, B, C, D, etc. on the ground as shown in Fig. 49.1.
- Set up the plane table at the station A. Select the point 'a' on the drawing paper such that the whole traverse can easily be accommodated on the paper.
- Level the table and centre it on the station A such that when the plumbing fork is set at 'a', the plumb bob is exactly over the peg at A and clamp the table.
- Mark the direction of the magnetic meridian on the drawing paper using the trough compass.
- Pivot the alidade on the plotted point 'a'. Sight the ranging rod at station B through the alidade and draw a line 'ab' along the fiducial edge of the alidade.
- Measure the distance AB in the field with a tape or a chain and select suitable scale.
- Lay off the length 'ab' equal to the distance AB in the field, and thus fix the position of the point 'b' on the line ab'.
- Locate the surrounding details on the line AB by using the method of radiation.
- The radial lines are drawn from the traverse station A to the details using the alidade.
- The distance of distance of each detail is measured in the field and laid off on the drawing paper.
- For distant objects and details the method intersection may be used. The radial lines are drawn from the traverse station A to the objects but the distances are not measured.
- The objects are located by drawing radial lines from the other traverse stations and finding out the points of intersection of the lines.
- Shift the plane table and set it up at the station B. Level and centre it over B and orient it by back sighting on A.
- With the alidade pivoted on 'b', sight the ranging rod at the forward station C, and draw a line along the fiducial edge of the alidade.
- Measure the distance BC in the field and set off the length bc on the paper and thus fix the point 'c'.
- Locate the surrounding details on the traverse line BC using the method of radiation or intersection, as before.
- Repeat the above process at all other stations.

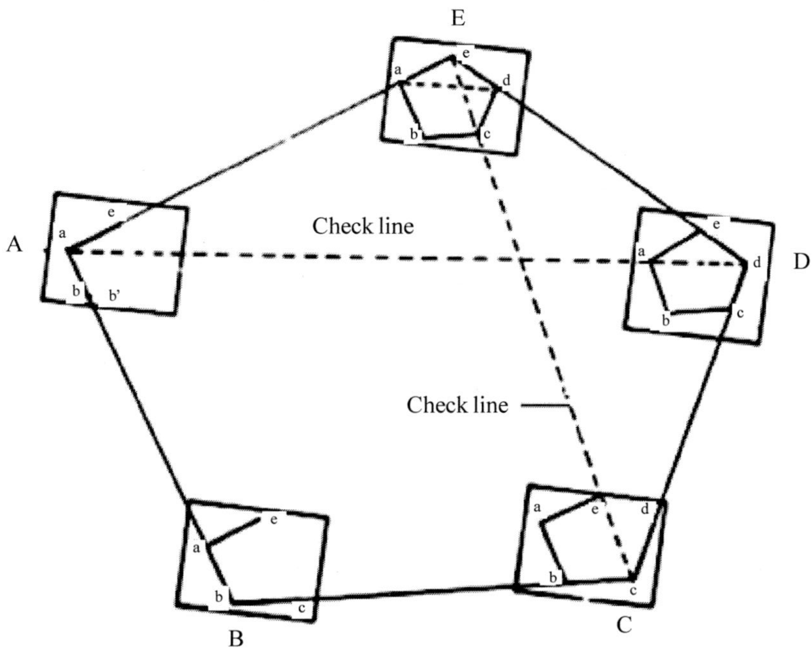

Fig.49.1. Schematic diagram of plane table traversing method

Check lines

To check the accuracy of the plane table traverse, few check lines are also taken. At any station, the check lines can be taken by sighting back to some preceding station. For example, the check line da is drawn from the station D to A when the plane table is occupying the station D. If the traverse is correct, the check line ad would pass through the plotted position a. Likewise, a check line ec is drawn from E to C.

Area calculation: Using the method of radiation or intersection, as discussed previous experiments.

Result

The area of given field by intersection method of plane table survey is

____

50

Studies on Leveling Instruments and Simple Leveling

Aim

- To study the levelling instruments
- To practice the simple leveling

Leveling

Leveling is a method of surveying used for determination of the difference of elevations or levels of various points on the surface of the earth. The elevation of a point is its vertical distance above or below a reference level, called datum. Most commonly used datum is the mean sea level (M.S.L). Leveling is required to determine the undulations of the earth's surface for topographic mapping. Leveling is needed for the design of highways, railways, canals, sewers, etc. Leveling is essential for the layout of construction projects, locating the excavation levels and for the control of various elevations in buildings, bridges, dams etc. The results of the leveling can be used to determine the catchment area, volume of the reservoir and the area submerged by a reservoir.

Dumpy level

The instrument, which is used for determine the difference in elevations of the points is called level. Dumpy is mainly used for leveling works. The dumpy level consists of a telescope, diaphragm, level tube, eyepiece, inclined mirror, leveling head and tripod (Fig.50.1).

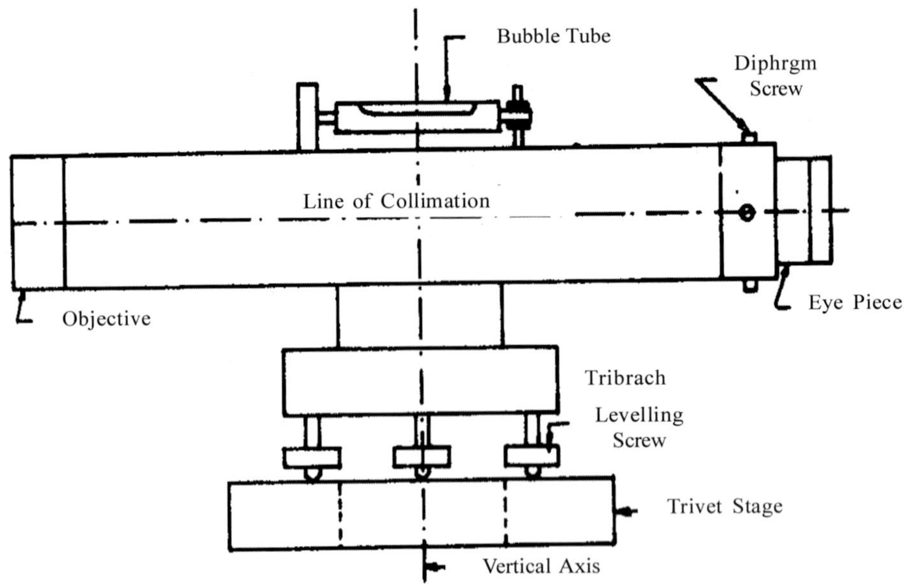

Fig.50.1. Schematic diagram of dumpy level

Telescope

The most important part of the dumpy level is telescope, which may be either internal focusing type or external focusing type. The telescope of the dumpy level is generally of the internal focusing type. The function of the telescope is to bring the image of the object formed by the objective exactly in the plane of cross hairs. The telescope consists of objective and eye-piece. These are mounted on two different tubes so that one can slide inside the other. There is a rack and pinion arrangement, which is operated by the focusing screw provided on the outer tube. By turning the focusing screw, the objective is moved in or moved out relative to the cross hairs (Fig.50.2). When the focusing is achieved, a clear image of the objective is formed in the plane of the cross hairs. Before focusing the objective, the eye-piece is adjusted so that the cross-hairs come into sharp focus. The eye-piece is screwed in or out of the main telescope by turning the eye-piece ring. For focusing the eye-piece, a piece of white paper is kept near the objective. The eye-piece ring is turned till the cross-hairs appear clear and distinct.

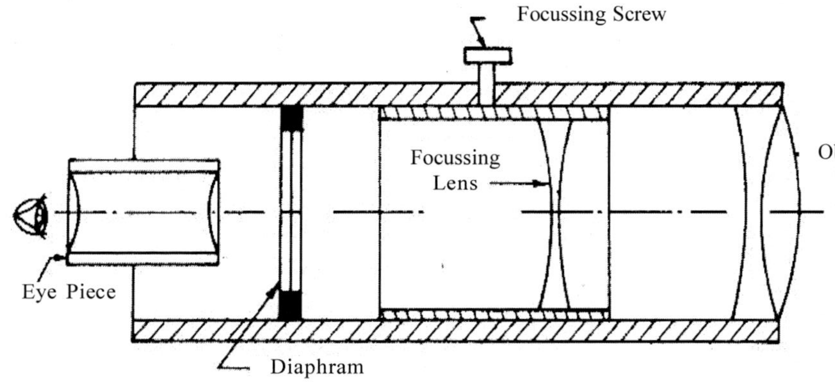

Fig.50.2. Telescope

ii) Diaphragm

The surveying telescopes are designed such that the image of the object is always formed in a fixed plane when the instrument has been focused. The fixed plane at which the image is formed is always made to lie in the plane of the cross-hairs in a ring called the diaphragm (Fig.50.3). The cross-hairs of the diaphragm normally have a horizontal line and vertical line. The horizontal hair is used to read the leveling staff and the vertical hair is used to check the verticality of the staff. The line joining the point of intersection of the cross-hairs and the optical center of the objective is called the line of sight or the line of collimation. The diaphragm also provided with two additional horizontal hairs, one above and other below the central horizontal hair. These hairs are called stadia hairs, and are used for computing the horizontal and vertical distances from the instrument. When the instrument is in adjustment, the line of sight is perpendicular to the vertical axis.

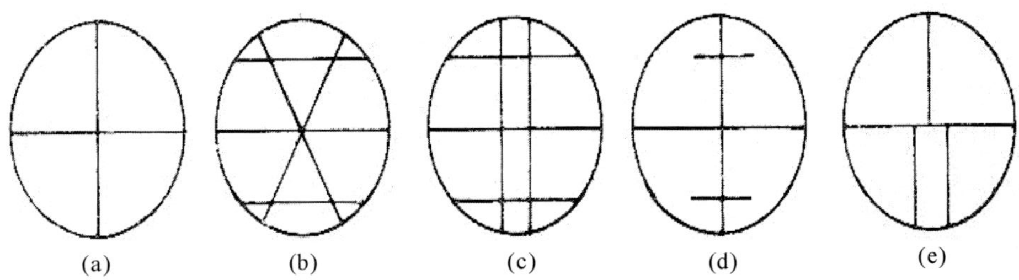

Fig.50.3. Various arrangements of cross hairs in diaphragm

iii) Level tube

A sensitive level tube is fitted on the top of the telescope or on its one side. It is used to leveling up the leveling instrument. It consists of a glass tube partially filled with a liquid which is non freezing and low viscosity (Fig.50.4). Purified synthetic alcohol is commonly used. The space above the liquid surface is occupied by vapors called the bubble. The tube has graduations on the upper surface and is spaced at 2 mm interval and is symmetrically center of the tube. On account of the action of gravity on the liquid, the bubble moves to the highest point of the tube. When the bubble is centered in the tube, the axis of the telescope is horizontal. The level tube is fixed in a cylindrical metal tube with plaster of Paris and is kept open at the top so that the bubble and graduations can be seen.

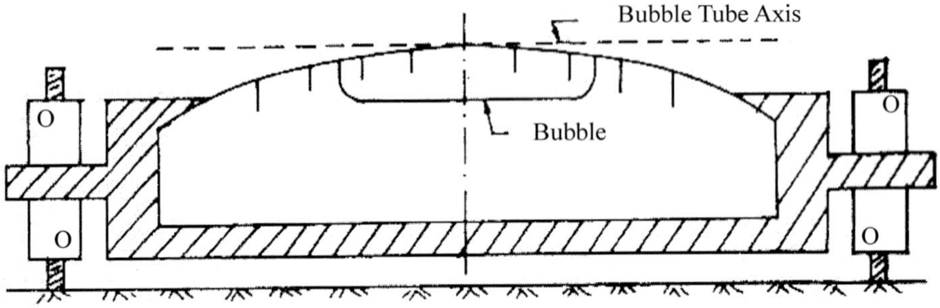

Fig.50.4. Level tube

iv) Eye-piece

The eye-piece is fitted on the telescope at the near end, whereas at the far end, there is objective. The eye-piece is a sort of microscope used for viewing the enlarged image of the object and the cross-hairs (Fig.50.5). The Ramsden eye-piece with plano-convex lenses is commonly used.

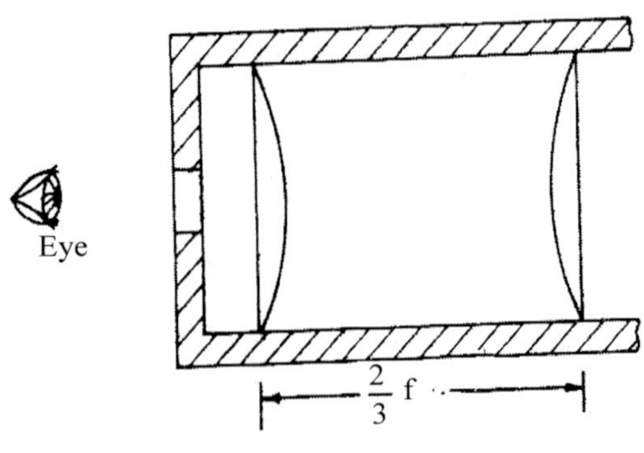

Fig.50.5. Eye piece

v) Inclined mirror

An inclined mirror is attached to the level tube to enable the observer to view the bubble from the eye-piece end of the telescope without moving round the instrument.

vi) Leveling head

The leveling head consists of two parallel plates held apart by three leveling screws. The upper parallel plate is called tribrach and the lower plate, known as trivet stage. The lower plate is screwed on the top of a tripod when the instrument is to be used. The telescope is fixed on a vertical spindle. The telescope and vertical spindle are cast as one piece. The spindle revolves in the socket of the leveling head. The telescope can be rotated in the horizontal plane about its vertical axis.

vii) Tripod

The level is fitted to the tripod when using the levels in the field. The tripod consists of a head and three legs with pointed metal shoes. The tripod head has an external screw to which the lower plate (trivet stage) of the level is screwed. Tripods are generally made of wood or aluminium. The legs of the tripod may have a constant length or may be extensible. The tripod shoes should be pressed into the ground to provide lateral stability to the legs.

viii) Leveling staff

Leveling staff is a graduated rod of rectangular cross-section. It is a giant size ruler used for the measurement of vertical distances. The leveling staff is kept vertical at the point and the reading of the staff is taken from the leveling instrument. Telescopic staff consists of three pieces which can be extended to the full length of 5 m. The upper piece is a solid piece having a length of 1.5 m. The central and lower pieces are hollow from inside having a length of 1.5 m and 2 m respectively. The upper piece can slide into central piece and the central piece can slide into lower piece. The overall length of the staff thus becomes 2 m when the staff is not in use. There is brass flat spring clamps fixed at back of the lower end of each piece for kept in position when the staff is extended to the full length.

Graduations are marked on the staff from the bottom of the brass cap. The bottom of the staff represents zero reading (Fig.50.6). Each meter is divided into 200 divisions. Divisions are painted alternatively black & white. The thickness of each division is 5 mm (= 0.005 m). The distance of 0.100 m is marked in line with the top of the 20th division from the bottom. The graduations are marked inverted so that when viewed through a telescope they appear erect.

Leveling procedure

i) Tripod leg adjustment

Erect the tripod at desired point and convenient height for sighting. Spread the tripod legs well apart Bring all foot screws in the centre of their run.

ii) Fixing the instrument

The instrument is fixed on the tripod by turning round the lower part and screw the instrument firmly. Fix any two legs firmly into the ground by pressing them and move the third leg to the right or left until the main bubble is approximately in the centre. Then move it in or out until the bubble of the cross level is approximately in the centre.

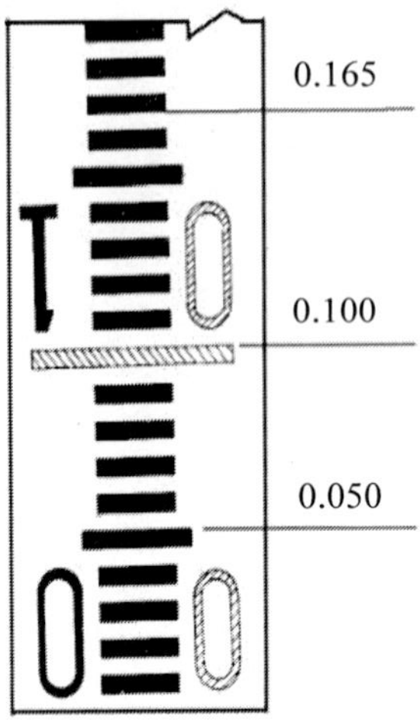

Fig.50.6. Graduations in leveling staff

iii) Leveling up

Place the telescope parallel to a pair of foot screws and bring the bubble to the centre of its run by turning these screws equally either both in wards or both outwards. Turn the telescope to 90° so that it lies over the third foot screw and centre the bubble by turning this screw. Repeat the operations until the bubble remains in the centre of its run in both positions.

iv) Focusing the eye piece

Remove the lid from the object glass and move the eye piece in and out until the cross hairs on the diaphragm are seen distinctly.

v) Focusing the object glass

Direct the telescope towards the staff and on looking through the eye-piece, bring the image of the staff between two vertical hairs of the diaphragm by lightly tapping the telescope. Adjust the objective by turning the focusing screw until the parallax error is eliminated.

vi) Focusing

Focus the eye-piece by holding a white paper in front of the telescope and move the eye- piece in and out until the cross hairs appear distinct and clear. Then focus the object glass by directing the telescope towards the object and turn the focusing screw until the image appears clear and sharp. By focusing, the focus of the objective and that of eye-piece coincide with cross hair of the diaphragm.

(vii) Holding the staff

While reading the staff, care should be taken in holding the staff in vertical. The staff man stands behind the staff, heels together, with the heel of the staff between his toes and holds it between the palms of his hands at the height of his face in a vertical position.

viii) Reading the staff

Direct the telescope towards the staff held vertically on the station after the instrument is leveled. Bring the staff between the two vertical hairs and use the portion of the horizontal cross hair between them in reading the staff. Bubble should be in the centre of its run while reading the staff.

ix) Booking staff readings in field book

- The readings should be entered in the respective columns and in the order of their observation.
- The first reading is obviously a back sight and should be entered in that column.
- A remark should be made in the remarks column describing weather back sight (B.S) is taken on a bench mark (B.M) and its value should be noted.
- If more than one reading is taken from the same position of the instrument, all the subsequent readings should be recorded in the intermediate sight column (I.S).
- The last reading is recorded in foresight column (F.S).
- The R.L of the plane of collimation should be written in the same horizontal line opposite to B.S.

Simple leveling

- Simple levelling is used for determination of the difference of levels of two points which are quite close to each other so that only one setting of the instrument is sufficient.
- The levelling instrument is set up at a suitable place midway between the two points. The instrument need not lie on the line joining the two points.
- Fig.50.7 shows the instrument set up at L between the points P and Q. The point P is of known elevation (R.L = 100.000). The elevation of the point Q is required.

- After doing the adjustments, take a back sight on the staff held at P. The back sight (B.S) reading indicates by how much vertical distance the line of sight is above the point P.
- After taking the back sight, the telescope is turned towards the point Q.
- The line of sight moves in the horizontal plane when the telescope is rotated, and therefore, its level remains the same.
- Take a fore sight (F.S) on the staff held at Q. The foresight reading will indicate the vertical distance of the line of sight above the point Q.

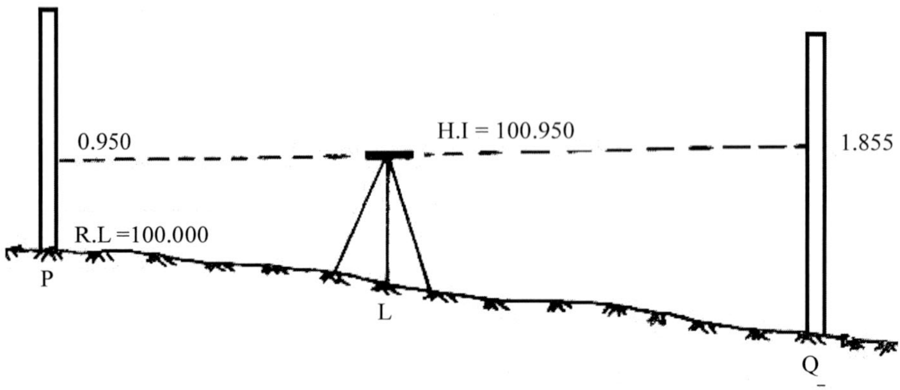

Fig.50.7. Diagram showing simple leveling

Calculation

Bench mark is known elevation, say, 100.000

R.L of P	= 100.000
Height of instrument (H.I)	= R.L of P + B.S
	= 100.000 + 0.950 = 100.950
R.L of Q	= H.I - F.S
	= 100.950 - 1.855 = 99.095
Elevation	= R.L of P – R.L of Q
	= 100.000 - 99.095 = 0.905 m

51

Determination of The Elevation of Different Points With Dumpy Level

Aim

- To determine the reduced level of various points
- To find out difference of elevations of two points

Materials required

- Dumpy level
- Levelling staff

Theory

Determination of the elevations and difference of elevations of two points is called differential leveling. A number of settings of the instrument are required in differential leveling. The differential leveling is also required in profile leveling, cross sectioning, contouring and other leveling operations.

Leveling procedure

i) Tripod leg adjustment

Erect the tripod at desired point and convenient height for sighting. Spread the tripod legs well apart and bring all foot screws in the centre of their run.

ii) Fixing the instrument

The instrument is fixed on the tripod by turning round the lower part and screw the instrument firmly. Fix any two legs firmly into the ground by pressing them and move the third leg to the right or left until the main bubble is approximately in the centre. Then move it in or out until the bubble of the cross level is approximately in the centre.

iii) Leveling up

Place the telescope parallel to a pair of foot screws and bring the bubble to the centre of its run by turning these screws equally either both in wards or both outwards. Turn the telescope to 90° so that it lies over the third foot screw and centre the bubble by turning this screw. Repeat the operations until the bubble remains in the centre of its run in both positions.

iv) Elimination of parallax

Parallax is a condition arising when the image formed by objective is not the plane of cross-hairs. Unless the parallax is eliminated accurate sighting is not possible. Parallax can be eliminated by (i) focusing the eye-piece and (ii) focusing the objective.

v) Focusing the eye piece

Remove the lid from the object glass and move the eye piece in and out until the cross hairs on the diaphragm are seen distinctly.

vi) Focusing the object glass

Direct the telescope towards the staff and on looking through the eye-piece, bring the image of the staff between two vertical hairs of the diaphragm by lightly tapping the telescope. Adjust the objective by turning the focusing screw until the parallax error is eliminated.

vii) Focusing

Focus the eye-piece by holding a white paper in front of the telescope and move the eye- piece in and out until the cross hairs appear distinct and clear. Then focus the object glass by directing the telescope towards the object and turn the focusing screw until the image appears clear and sharp. By focusing, the focus of the objective and that of eye-piece coincide with cross hair of the diaphragm.

viii) Holding the staff

While reading the staff, care should be taken in holding the staff in vertical. The staff man stands behind the staff, heels together, with the heel of the staff between his toes and holds it between the palms of his hands at the height of his face in a vertical position.

ix) Reading the staff

Direct the telescope towards the staff held vertically on the station after the instrument is leveled. Bring the staff between the two vertical hairs and use the portion of the horizontal cross hair between them in reading the staff. Bubble should be in the centre of its run while reading the staff.

Field measurement procedure

- The instrument should be set-up at a convenient point (L_1). After completing the levelling procedure, take back sight (B.S) reading on temporary bench mark.
- The levelling staff should be held vertically on temporary bench mark that should be selected in the survey field.
- The back sight (BS) reading should be entered in the BS column. A remark should be made in the remarks column describing weather back sight (BS) is taken on a benchmark (BM) and its value should be noted.
- The staff man then moves to another point 'B' there the staff should be held towards the instrument and take reading.
- Similarly, numbers of readings known as intermediate sights (IS) should be taken on C, D, E, F, G, H. The last reading said to be fore sight (F.S) and in this point (H) the staff position should not be changed.
- The IS and FS reading should be entered in their respective columns.
- Then the instrument should be shifted to another convenient point (L_2). Levelling and focusing should be made.
- Back sight (B.S) reading is taken on the staff held at point H. It may be emphasized that the staff is not moved when the instrument is being shifted.
- Only the face of the staff is turned so that it faces the instrument at L_2.
- The staff man then moves to other points I, J, K, L, M, N, O. These readings are intermediate sights (I.S) and last reading is fore sight (F.S).
- The I.S and F.S reading should be entered in their respective columns of the field book.
- Find out the reduced level (R.L) of all the points using height of instrument (H.I.) method.

Data tabulation

Table 51.1. Collimation or height of instrument (H.I.) method

Station	Readings, m			H.I, m	R.L, m	Remarks
	B.S	I.S	F.S			
A						Benchmark = 100.000 RL of BM = 100.000
B						
C						
D						
E						
F						
G						
H						
I						
J						
K						
L						
M						
N						
O						
	$\sum$ B.S		$\sum$ F.S			

Calculation of reduced level

Example

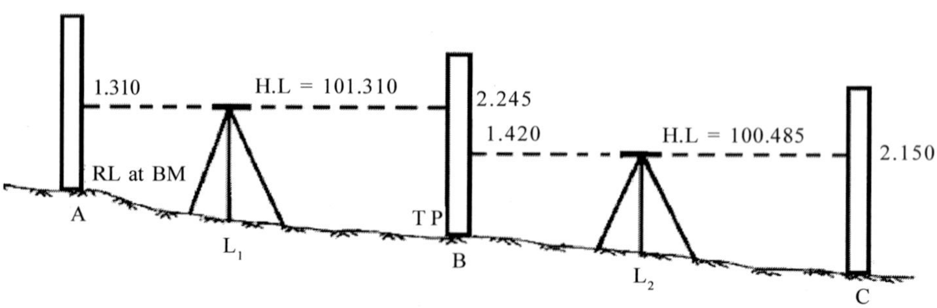

Fig.51.1. Diagram shows the differential leveling

Table 51.2. Height of instrument method

Station(1)	B.S(2)	I.S(3)	F.S(4)	H.I(5)	R.L(6)	Remark(7)
A	1.310			101.310	100.000	B.M
B	1.420		2.245	100.485	99.065	T.P
C			2.150		98.335	
Sum	2.730		4.395			

R.L of A = 100.000

H.I at L_1 = R.L of A + B.S

= 100.00 + 1.310

= 101.310

R.L of B = H.I at L_1 – F.S

= 101.310 – 2.245

= 99.065

H.I at L_2 = R.L of B + B.S

= 99.065 + 1.420

= 100.485

R.L of C = H.I at L_2 - F.S

= 100.485 – 2.150

= 98.335

The difference in elevation between station A and station C is

R.I of A – R.L of C

100.000 - 98.335

= 1.665 m

Arithmetic check

Sum of B.S – Sum of F.S = Last R.L – First R.L

2.730 – 4.395 = 98.335 – 100.000

-1.665 = - 1.665 (O.K)

Result

The difference of elevations of two points is _________

52

Profile Levelling Determination of Reduced Levels and Plotting of the Road Profile

Aim

- To determine the reduced levels on various points on the road
- To plot the profile of the road

Materials required

- Dumpy level
- Levelling staff

Theory

Profile leveling is the process of leveling along a fixed line to determine the elevations of the ground surface along that line. The fixed line is generally the centre line of a highway, railway, canal, sewer, etc. The profile leveling is done to determine the undulations of the ground surface. Cross-sectioning is required to determine the levels across the line along which profile leveling is done. Profile leveling should be commenced from a bench mark and should end at a bench mark, as far as possible. When the ground surface does not have many undulations, the intermediate points are fixed at a uniform interval. If the ground surface is irregular and there are abrupt changes in the ground surface, the interval may be non-uniform.

Leveling procedure

i) Tripod leg adjustment

Erect the tripod at desired point and convenient height for sighting. Spread the tripod legs well apart and bring all foot screws in the centre of their run.

ii) Fixing the instrument

Release the clamp screw of the instrument and hold the instrument in the right hand and fix it on the tripod by turning round the lower part with left hand and screw the instrument firmly.

Fix any two tripod legs firmly into the ground by pressing them with hand and move the third leg to the right or left until the main bubble is approximately in the centre. Then move the third leg in or out until the bubble of the cross-level is approximately in the centre and fix it firmly when the bubbles are approximately in the centres of their run.

iii) Leveling up

Place the telescope parallel to a pair of foot screws and bring the bubble to the centre of its run by turning these screws equally either both in wards or both outwards. Turn the telescope to 90° so that it lies over the third foot screw and centre the bubble by turning this screw. Repeat the operations until the bubble remains in the centre of its run in both the positions. Turn the telescope through 180° and check whether the bubble remains central

iv) Elimination of parallax

Parallax is a condition arising when the image formed by objective is not the plane of cross-hairs. Unless the parallax is eliminated accurate sighting is not possible. Parallax can be eliminated by (i) focusing the eye-piece and (ii) focusing the objective.

v) Focusing the eye piece

Remove the lid from the object glass and move the eye piece in and out until the cross hairs on the diaphragm are seen distinctly.

vi) Focusing the object glass

Direct the telescope towards the staff and on looking through the eye-piece, bring the image of the staff between two vertical hairs of the diaphragm by lightly tapping the telescope. Adjust the objective by turning the focusing screw until the parallax error is eliminated.

vii) Focusing

Focus the eye-piece by holding a white paper in front of the telescope and move the eye- piece in and out until the cross hairs appear distinct and clear. Then focus the object glass by directing the telescope towards the object and turn the focusing screw until the image appears clear and sharp. By focusing, the focus of the objective and that of eye-piece coincide with cross hair of the diaphragm.

viii) Holding the staff

While reading the staff, care should be taken in holding the staff in vertical. The staff man stands behind the staff, heels together, with the heel of the staff between his toes and holds it between the palms of his hands at the height of his face in a vertical position.

ix) Reading the staff

Direct the telescope towards the staff held vertically on the station and turn the focusing screw until a clear image of graduations (engraved in staff) formed in the plane of the cross hairs. Bubble should be in the centre of its run while reading the staff.

Field measurement procedure

- The instrument should be set-up at a convenient point (L_1). Hold the staff vertically over the temporary bench mark that should be selected in the survey field.
- After completing the levelling procedure, observe the staff reading on temporary bench mark. It is called back sight (B.S) reading.
- The back sight (BS) reading should be entered in the B.S column of the field book. A remark should be made in the remarks column describing weather back sight (B.S) is taken on a bench mark (B.M) and its value should be noted.
- The staff man then moves to another point 'B' there the staff should be held towards the instrument and observe staff reading. It is called intermediate sights (I.S)
- Similarly, numbers of staff readings known as intermediate sights (IS) should be observed on the intermediate points C, D, E, F, G, H, etc.
- Hold the staff on the last point and observe staff reading. The last reading said to be fore sight (F.S).
- The I.S and F.S reading should be entered in their respective columns of the field book.
- Find out the reduced level (R.L) of all the points using height of instrument (H.I.) method.

Data tabulation

Table 52.1. Collimation or height of instrument (H.I.) method

Station	Readings, m			H.I, m	R.L, m	Remarks
	B.S	I.S	F.S			
A						Bench mark = 100. 000R.L of BM = 100.000
B						
C						
D						
E						
F						
G						
H						
I						
J						
K						
L						
M						
N						
O						
	Ó B.S		Ó F.S			

Calculation of reduced level

Height of instrument (H.I) = R.L of A + B.S

Reduced level at I.S points, R.L = H.I - I.S

Reduced level at F.S point, R.L = H.I - F.S

Elevation = R.L at first point – R.L at last point

Plotting the profile

- After the reduced levels have been determined in the field, the profile is plotted. The profile is usually plotted on a specially prepared profile paper or a graph paper.
- The vertical scale is generally showing the undulations of the ground clearly.
- The vertical scale is conventionally taken 10 times the horizontal scale. Thus if the horizontal scale is 1:1000, the vertical scale would be 1 in 100.
- Select a suitable datum for plotting the profile. Draw a horizontal line to represent the datum line.

- Mark the chainages of the various points along the datum line. Draw perpendicular lines at each plotted chainage, and set off the respective levels at these perpendicular lines.
- Also write the R.L. of various points against these lines. Each point is thus plotted by the two co-ordinates, viz., its horizontal distance from the starting point and its vertical elevation above the datum.
- The plotted points are joined free hand to obtain the outline of the ground surface along the fixed line.
- Write the description of prominent features, if any, at the appropriate position on the section.

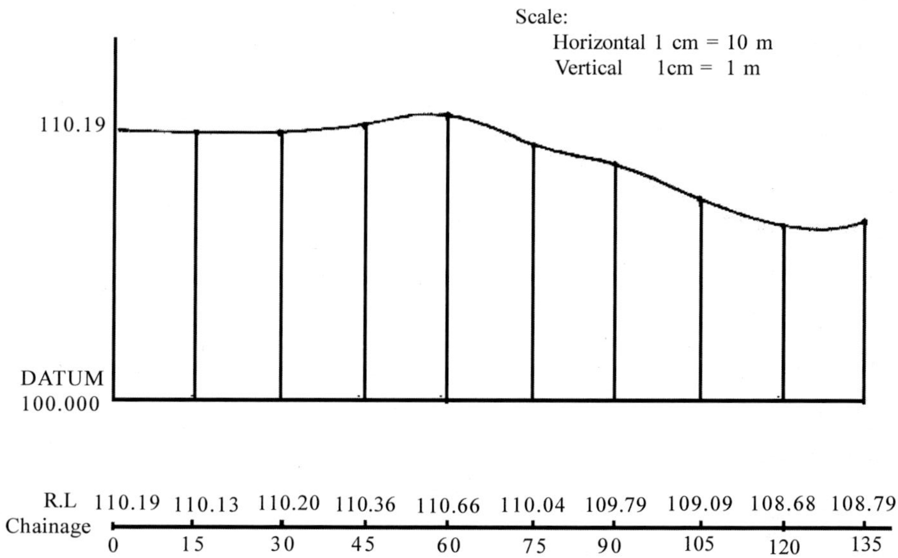

Longitudinal section

Fig.52.1. Plotting the profile of the field

53

Preparation of Contour Map by Direct Method

Aim

- To prepare a contour map by direct method using dumpy level.

Materials required

- Dumpy level
- Leveling staff
- Measuring tape
- Wooden pegs/chain pins
- Field book

Theory

For locating a point on the ground on a contour, its position and elevation in a horizontal plane should be known. In direct method, the contours to be plotted are located on the ground with a level by making various points on each contour, and these points are surveyed and plotted on plan. This method is more accurate but slow and tedious, and is suitable for contouring small areas where great accuracy is required. For locating the points of the same elevation, the instrument is set in such a commanding position from where maximum number of points can be covered. The bench mark is established near the site of survey. The height of the instrument (H.I) is determined after taking a back sight (B.S) on the established bench mark.

Procedure

- Establish a bench mark (BM) near the site of survey.
- Mark desired position of the points with pegs on the each contour line.
- The level is then set up in a commanding position (P) and accurately levelled.

- Take back sight (BS) reading on bench mark.
- The height of instrument (HI) is determined by adding BS and reduced level of the BM.
- The staff is held at an appropriate position of the points and looking through the level.
- The staff is moved up or down to get desired reading corresponding to a particular contour.
- The required staff readings (intermediate and fore sights) of various points are determined.
- In a similar manner, all the points are located and pegged.
- From the same set up of level, all the points are located.
- The level is then shifted to new position from where a new height of instrument is determined by taking a back sight on a previously surveyed point.
- New required staff readings are taken corresponding to this HI and various contours are determined.
- The process is repeated to locate all the points.
- Determine the reduced level (RL) of the points by subtracting IS and FS from HI.
- The RL of points should be plotted on a map and contours are drawn by joining the point of equal elevation.

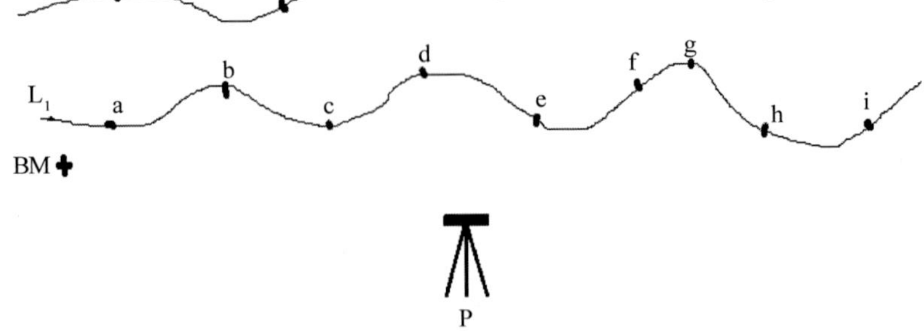

Fig.53.1. Direct method of locating contours

Observations

Survey points	BS	IS	FS	HI	RL	Remarks
						BM of RL = 100.000

Calculation of reduced level

Height of instrument (H.I) = R.L of A + B.S

Reduced level at I.S points, R.L = H.I - I.S

Reduced level at F.S point, R.L = H.I - F.S

Elevation = R.L at first point – R.L at last point

54

Determination of Contours by Grid Method

Aim

- To determine the contours by grid method using dumpy level.

Materials required

- Dumpy level
- Leveling staff
- Measuring tape
- Wooden pegs/chain pins
- Field book

Procedure

- The grid method is used when the area to be contoured is not very large and where the ground is not much undulating.
- In this method, the area to be surveyed is divided into a number of squares.
- The size of the square depends upon the nature of the ground and accuracy required.
- The sides of the square generally various from 5 m to 20 m.
- A base line is first ranged out for divide the field into squares.
- Then different parallel and perpendicular lines are laid out at fixed intervals to get the required square grids.
- The corners of the squares are pegged.
- The elevations of the corners are determined by using a level.
- To obtain the elevations of the grid, the level is set up in the middle/side of the area at a commanding position.

- The bench mark is established near the survey area.
- The H.I is determined by taking a back sight on the bench mark.
- The intermediate sights (I.S) are then taken on each of the grid points.
- The instrument is shifted to new position if needed and new height of instrument is determined.
- The process is repeated till completion work.
- The elevation of the grid points is determined.
- The points on various contours are located by interpolation from the elevations of the grid points.

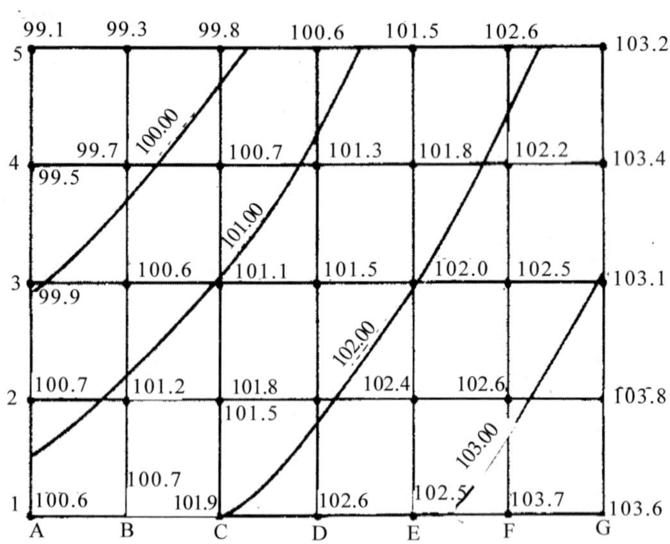

Fig.54.1. Sketch showing contour lines in grid method.

Observations

Survey points	BS	IS	FS	HI	RL	Remarks
						BM of RL = 100.000

Calculation of reduced level

Height of instrument (H.I) = R.L of A + B.S

Reduced level at I.S points, R.L = H.I - I.S

Reduced level at F.S point, R.L = H.I - F.S

Elevation = R.L at first point – R.L at last point

55

Determination of Contours by Cross-Section Method

Aim

- To determine the contours by cross-section method using dumpy level.

Materials required

- Dumpy level
- Leveling staff
- Measuring tape
- Wooden pegs/chain pins
- Field book

Theory

The cross-section method is generally used for the determination of contours along a fixed route, such as a canal, road, railway, etc. Cross sections are located on the ground at right angles to the fixed line of the route (Fig.55.1). The spacing of the cross-sections depends upon the nature of the ground, the contour interval and the purpose of contouring. The spacing should be kept small in ravines where the contours change directions abruptly.

Procedure

- AB is the fixed line of the route on the ground.
- Locate cross-sections CS1, CS2, CS3, CS4, CS5, CS6, etc at right angles to the fixed line.
- The spacing between the cross-sections are depending on the nature of ground.
- Select several guide points approximately equal distance on the cross-sections.

- The elevations of guide points on cross-section are determined using dumpy level.
- The staff readings taken at the points on breaks in the ground surface are recorded with their respective distances from the fixed line.
- From the elevations of the guide points, the contours are obtained by interpolation.

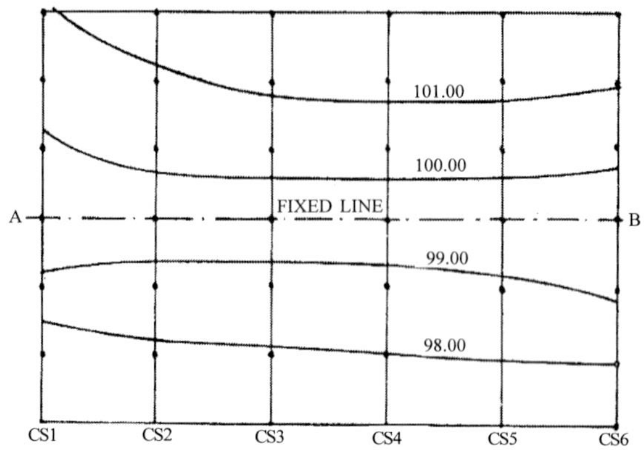

Fig.55.1. Sketch showing contour lines in cross-section method

Observations

Survey points	BS	IS	FS	HI	RL	Remarks
						BM of RL = 100.000

Calculation of reduced level

Height of instrument (H.I) = R.L of A + B.S

Reduced level at I.S points, R.L = H.I - I.S

Reduced level at F.S point, R.L = H.I - F.S

Elevation = R.L at first point – R.L at last point

56

Determination of Contours by Radial Lines Method

Aim

- To determine the contours by radial line method using dumpy level.

Materials required

- Dumpy level
- Leveling staff
- Measuring tape
- Wooden pegs/chain pins
- Field book

Procedure

- For contouring small hilly areas, radial lines are run from peak to cover area.
- The guide points are taken on the radial lines and their elevations are determined.
- The level is set up in a single point in the center can cover the entire area (Fig.56.1).
- Radial lines are ranged out from the common center point by using a compass
- The positions of these points are fixed by horizontal angles and bearings.
- Temporary bench marks are fixed both at the center and near the ends of the radial lines.
- The contour points are located and pegged by using the method explained earlier.
- The distances are measured along the radial lines.
- They are then plotted and the contours are obtained by joining all the corresponding points.

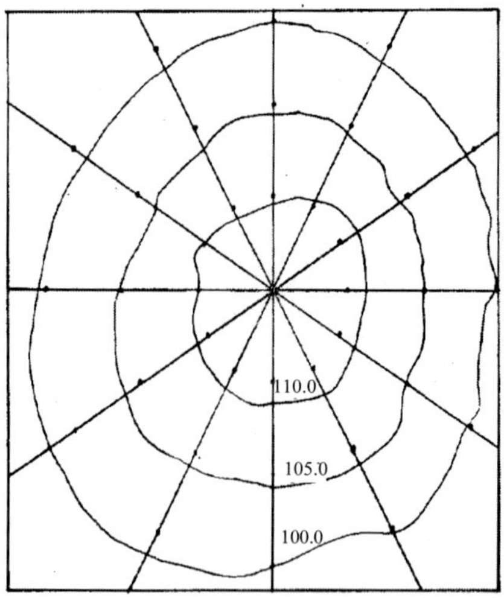

Fig.56.1. Sketch showing contouring in radial lines method

Observations

Survey points	BS	IS	FS	HI	RL	Remarks
						BM of RL = 100.000

Calculation of reduced level

Height of instrument (H.I) = R.L of A + B.S

Reduced level at I.S points, R.L = H.I - I.S

Reduced level at F.S point, R.L = H.I - F.S

Elevation = R.L at first point – R.L at last point

57

Measurement of Irrigation Water by Volumetric Method

Aim

- To measure irrigation water by volumetric method

Apparatus required

- Siphon tube
- Stop watch
- Measuring bucket

Theory

The knowledge of water measurement is highly essential to apply water to the crops precisely as per required quantity. The planning of cropping system, design of irrigation method and irrigation project is depends upon the available discharge. Most of the drainage and water supply systems are require discharge measurement. The measurement of discharge is also essential, if desire to charge the water rates to the farmers on volume basis. Irrigation water is conveyed either through open channels or pipes and knowing the quantity of water available is essential for irrigation water management.

Units of measurement

- In volumetric method, the water can be measured at rest position.
- In rest position, water is measured in volume and expressed in different units such as litres, cubic meter, hectare-centimeter, hectare-meter depending upon the magnitude and purpose of measurement.
- Large quantity of water present in a reservoir may be measured in cubic meter.
- Volume of water applied to a field can be suitably measured as hectare-centimeter.

- Cubic metre: Volume of water equal to that of a cube having 1 m long, 1 m wide and 1 m height.
- Hectare-centimetre: Volume of water required to cover an area of 1 hectare (= 10,000 m^2) up to a depth of 1 cm (1 ha-cm = 100 m^3 = 100,000 litres).
- Hectare -metre: A volume necessary to cover 1 hectare (10,000 m^2) up to a depth of 1 m (1 ha-m = 10,000 m^3 = 10 million litres).

Procedure

- This is the most simple and accurate method and can be used to measure small discharge flowing through a channel or a pipe.
- In this case, water is discharged through a siphon tube from irrigation channel
- The water is collected in a container of known volume which is placed below the channel level (Fig.57.1).
- Time to fill the container is noted with the help of a stop watch.

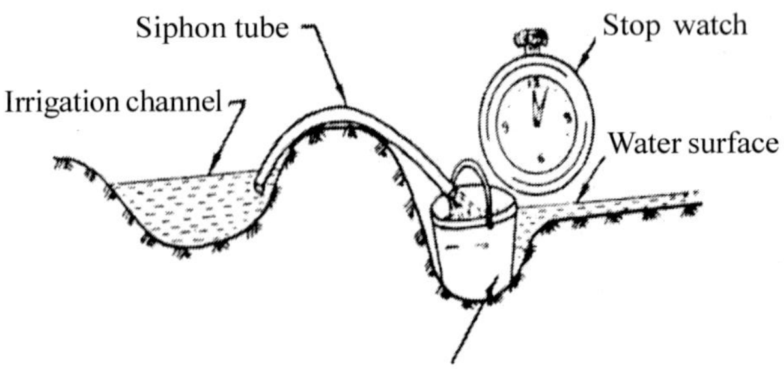

Fig. 57.1. Volumetric method of water measurement

Calculation

The rate of flow is measured by the following formula:

$$\text{Disch arg erate, L / S} = \frac{\text{Volume of container, liters}}{\text{Time required to fill the container, sec ond}}$$

Result

The discharge rate of water through a channel is __________

58

Measurement of Irrigation Water by Area-Velocity Method

Aim

- To measure irrigation water by area-velocity method by using (i) float and (ii) current meter

Apparatus required

- Float
- Current meter

Theory

The rate of flow of water in open channel can be determined by multiplying the cross sectional area of the flow section at right angles to the direction of flow by the average velocity of water. In area-velocity method, the water can be measured at motion. The water is measured in motion by means of rate of flow or discharge and is expressed in litre per second (lps), litre per minute (lpm), cubic metre per second (m^3/s), etc. Litre per second: A continuous flow amounting to 1 litre passing through a point each second. Cubic metre per second: A flow of water equivalent to a stream of 1 m long, 1 m wide and 1 m deep flowing at a velocity of 1 metre per second. In area-velocity method, the velocity of water is measured either by float or current-meter

Procedure

i) Determination of the area of the stream

- The area of the stream can be determined by the cross section is divided into several subsections.
- The depth of water (d) at each subsection at equal interval across the water flow is determined (Fig.58.1).
- The average depths (h) at each subsection also determined between the depths.

- The depth can be measured by different methods like sounding rods or sounding weights or echo depth sounder for accurate measurement.
- The cross sectional area can be determined by following formula:

$$A = W \times (h_1 + h_2 + h_3 + ... + h_{n-2} + h_{n-1} + h_n)$$

$$h_1 = \frac{d_0 + d_1}{2}$$

$$h_2 = \frac{d_1 + d_2}{2}$$

Where,

A = cross-sectional area of channel in which water flowing, m^2

w = common distance between the depths

h_1, h_2, h_3,...... are depth at the midpoint of each division

d_0, d_1, d_2, ... are the depths

- The cross-sectional area of the stream is measured at least in 3 places and the average is taken. The average velocity at each of these sub-sections is determined by floats or current meters.

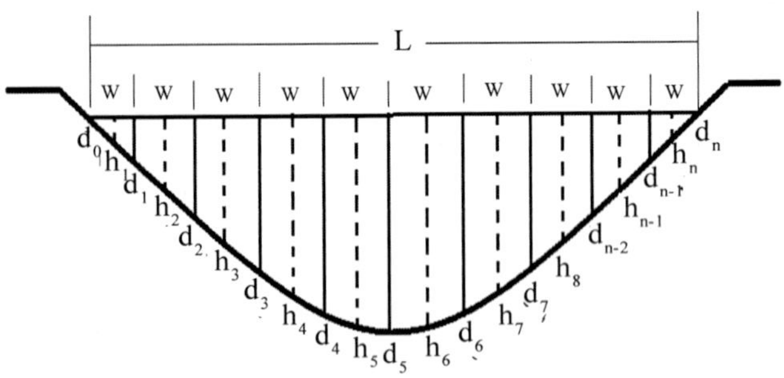

Fig.58.1. Stream section for area-velocity method

ii) Determination of velocity of water in stream

a) Float method

- This is one of the simplest methods and gives appropriate velocity of flow in a channel if no other means is available.
- Any object that flows in water but does not obstruct wind or get carried away by it, can be used as a float.
- A dry wood of about 1 cm^3 size serves the purpose best.

- A straight section of the channel of about 30 meter length is selected such that the cross-section is almost uniform through the length.
- The float is allowed to move from the upstream to the downstream end and time taken to move this 30 meter distance is measured.
- The velocity of flow of water can be determined by the following relationship:

$$\text{Velocity, m/s} = \frac{\text{Distance travelled by float, m}}{\text{Time taken, sec}}$$

- Several measurements are taken at every subdivision and find out average velocity.
- This is called surface velocity which is always higher than the average stream velocity.
- Therefore, the surface velocity is multiplied by a factor of 0.85 to get the average velocity.

b) Current meter method

- Two types of current meters viz. cup type and the propeller type are used for measurement of velocity in irrigation channels, streams or rivers.
- The cup type current meter (Fig.58.2) consists of a wheel having several cups and the propeller type consists of vanes or propeller attached to a rod.
- When the meter is placed in flowing water, the velocity of water rotates the wheel or the vane.
- In case of shallow streams, the current meter is suspended from a rod and for measurement in deep water, a cable is used to suspend the meter.
- In large rivers, a boat is used to take readings inside the river. Sometimes the current meter is suspended from a bridge if available.
- As the wheel rotates, the revolution is recorded in a counter.
- A calibration chart or a graph gives the relationship between the revolution per minute and the velocity.
- The average velocity is obtained by taking the average of the velocities at 0.8 and 0.2 of the depth from the top.
- In case of shallow stream, if it is not possible to take the reading at 0.8 of the depth from the surface, only one reading may be taken at 0.6 of the depth from the surface.
- For discharge calculation the entire cross section is divided into several subsections and the accuracy of discharge measurement increases with the increase in the number of segments.

iii) Discharge rate

The discharge rate is measured by the average velocity is multiplied by the average cross-sectional area:

$Q = A \times V$

Where,

Q = discharge, m^3/s

A = cross-sectional area, m^2

V = average velocity of flow, m/s

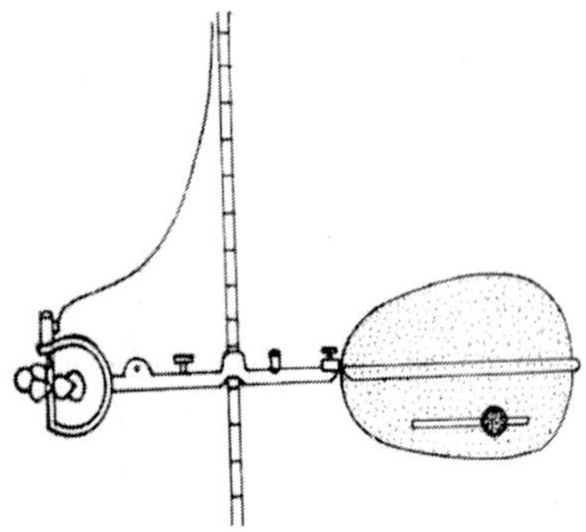

Fig.58.2. Cup type current meter

Result

(i) The discharge rate of irrigation water using float is ____________

(ii) The discharge rate of irrigation water using current meter is ______________

59

Measurement of Irrigation Water by Using Weirs

Aim

- To measure the irrigation water by using weirs

a) Weirs

A weir is a notch or opening of some definite shape installed on a channel or a stream through which waterfalls (Fig.59.1). Materials like wood, concrete, mild steel, rigid PVC etc. can be used for construction of a weir. Mainly a cut on a sheet metal is used for fabrication of weirs, to be installed at various locations in a channel. Advantages of weirs are (i) capable of accurately measuring a wide range of flows (ii) easy to construct, portable and adjustable (iii) tends to provide more accurate discharge rating than flumes and orifices.

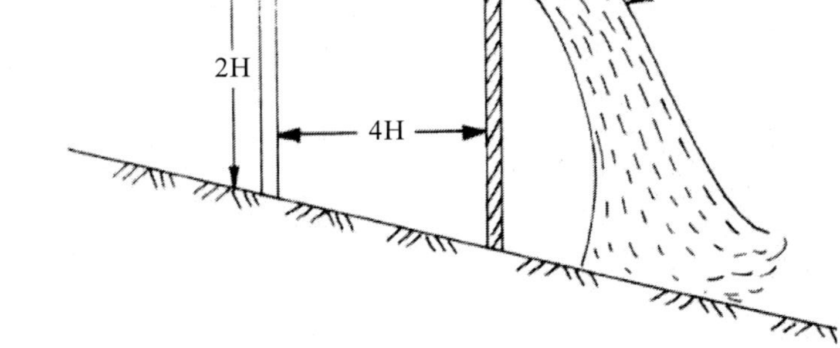

Fig.59.1. Weir parts

Weir terminologies

Weir pond: The portion of channel immediately upstream from the weir

Weir crest: The sheet of water is flowing over its thin edge with minimum contact area. A portion of weir in touch with water is called weir crest.

Nappe: The sheet of water flowing over the weir is called the nappe. The top surface is the upper nappe and the bottom surface is the lower nappe.

Head: The depth of flow over the crest is known as head (H). It is measured at a minimum distance of 4 H upstream from the crest.

End contraction: The horizontal distance from the end of the crest to the side of the channel is known as end contraction. If both ends of the crest are away from the sides of the channel, then there are two end contractions. If the length of the crest is same as the width of the channel, then there is no end contraction.

Free flow condition: If the surface of water downstream is below the crest level so that the nappe is surrounded by air then the weir said to have free flow condition.

Submerged flow condition: If the downstream water level is higher than the weir crest level then it is a submerged flow condition.

Classification of weirs

Weirs can be classified as (i) broad crested and (ii) sharp crested weirs. The sharp crested weir is mostly used for measurement of irrigation water.

Broad-crested weir:

A weir having a horizontal crest sufficiently long in the direction of flow so that the nappe will be supported and hydrostatic pressures will be fully developed for at a short distance. When the crest is broad, the streamlines become parallel to the crest invert and the pressure distribution above the crest is hydrostatic. The broad-crest weir is shown in Fig.59.2.

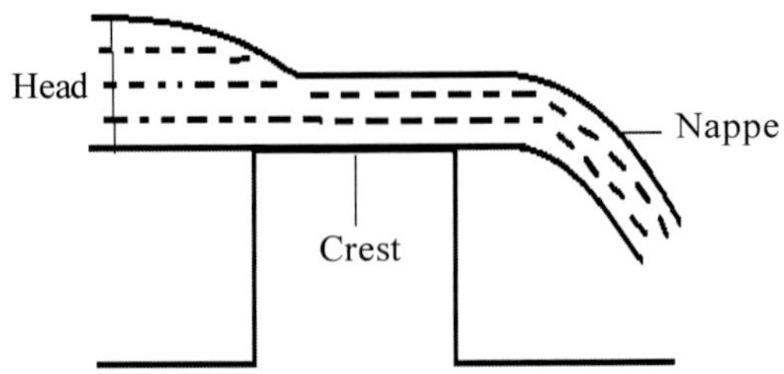

Fig.59.2. Broad-crested weir

Sharp crested weir

A weir having thin edged crest such that the over flowing sheet of water has the minimum surface contact with the crest. A sharp-crested weir allows the water to fall cleanly away from the weir. The sharp crested weir is mostly used for measurement of irrigation water. They are generally of three types depending upon the shape of notch. These are

a) Rectangular weir

b) Trapezoidal weir or Cipoletti weir

c) Triangular weir or V-Notch weir

Discharge formula for weirs

$$Q = C \times L \times H^{m}$$

Where,

Q = Discharge rate, lit/s

L = Length of the weir crest, cm

H = Head on the crest, cm

m = exponent which depends upon the type of weir opening

$(C = \frac{2}{3}\sqrt{2g} = 0.0184)$

C = Discharge coefficient of weir (depends on nature of the crest and approach condition)

Rectangular weir

The rectangular weir is used to measure high discharge. Its crest is horizontal and the sides are vertical. The discharge through the weir is directly related to the water depth (H). The head is affected by the condition of the crest, contraction, velocity of approaching stream and elevation of the water surface downstream from the weir. In case, the crest length is same as that of the channel width, it is known as suppressed weir, otherwise contracted weir.

Suppressed rectangular weir

A rectangular weir whose notch or opening sides are coincident with the sides of the approach channel, also rectangular, which extend unchanged downstream from the weir. It is the lateral flow contraction that is suppressed. Fig.59.3 shows suppressed rectangular weir.

Discharge for rectangular weirs

$$Q = 0.0184(L - 0.1nH)H^{3/2}$$

where, n is the number of end contractions

For suppressed rectangular weir, number of end contraction is zero:

$$Q = 0.0184 \times L \times H^{3/2}$$

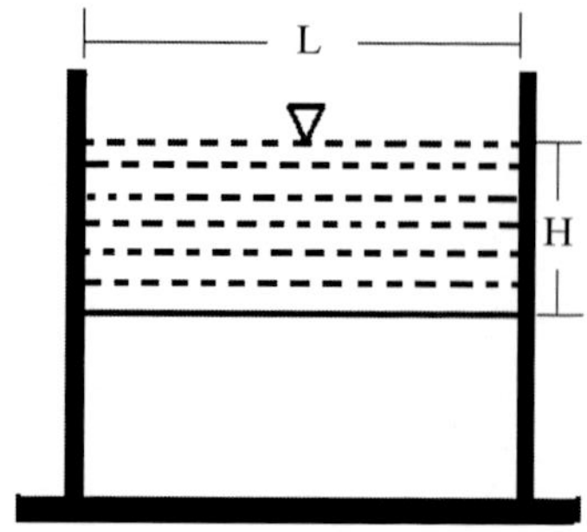

Fig.59.3. Suppressed rectangular weir

Contracted rectangular weir

The sides and crest of a weir are far away from the sides and bottom of the approach channel (Fig.59.4). The nappe will fully contract laterally at the ends and vertically at the crest of the weir. It is also called an unsuppressed weir.

Discharge for one side contraction

$$Q = 0.0184(L - 0.1H)H^{3/2}$$

Discharge for both side contraction

$$Q = 0.0184(L - 0.2H)H^{3/2}$$

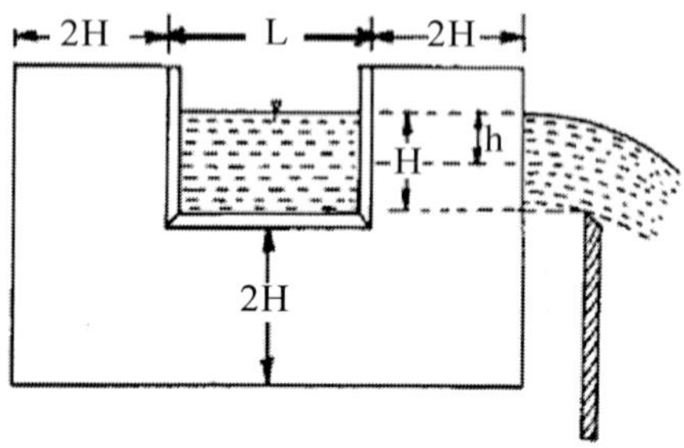

Fig.59.4. Contracted rectangular weir

Trapezoidal weir

The trapezoidal weir (Fig.59.5) has a horizontal crest and the slides slope outward to give the notch a trapezoidal cross-section Commonly a side slope of 1 horizontal to 4 vertical is used and it is named as cipoletti weir. It does not require any correction for end contractions and is used for measurement of medium discharges. Normally it is a sharp crested weir.

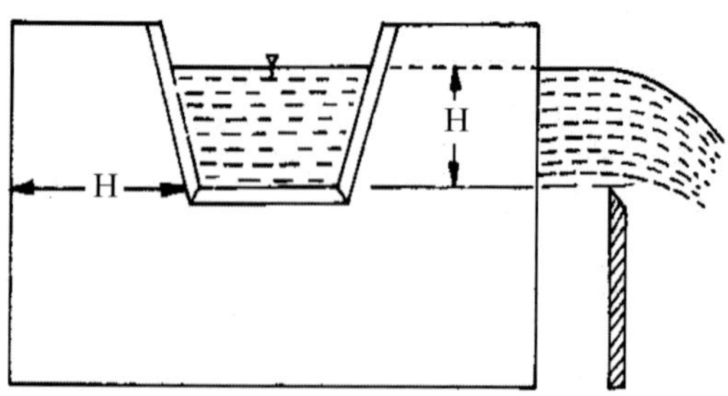

Fig.59.5. Trapezoidal weir

Discharge for cipoletti weir

$$Q = 0.0186 \times L \times H^{3/2}$$

Triangular or V-notch weir

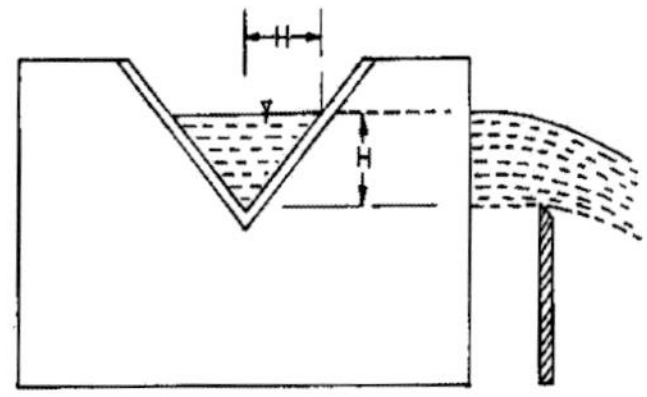

Fig.59.6. V-Notch weir

It is used to measure small to medium discharges. The 90° V-notch weir (Fig.59.6) is used to measure low discharges accurately. Thus the sides of the notch make an angle 45 ° with the vertical which gives a slope of 1 horizontal to 1 vertical.

Discharge for V-notch or triangular weir

$$Q = 0.0138\, H^{5/2}$$

Example 1: A rectangular weir of crest length 30 cm is installed at the centre of a rectangular channel of 50 cm width. The height of water above the weir crest is 10 cm. Calculate the discharge.

Solution: Data given: H = 10 cm; L = 30 cm Number of contractions, n = 2

$$Q = 0.0184(L - 0.1nH)H^{3/2}$$

$$Q = 0.0184 \times (30 - 0.1 \times 2 \times 10) \times 10^{3/2} = 16.29\,\text{lps}$$

Example 2: A cipoletti weir of 50 cm crest length is installed in a straight channel and the head over the crest of the weir is 15 cm. Calculate the discharge.

Solution: Data given: L = 50 cm; H = 15 cm

$$Q = 0.0186 \times L \times H^{3/2}$$

$$Q = 0.0186 \times 50 \times 15^{3/2} = 54.03\,\text{lps}$$

Problem 3: Water flows through a contracted rectangular weir 150 cm long to a depth of 40 cm, it then flows along a rectangular channel 180 cm wide and over a second weir which has length equal to the width of the channel. Determine the depth of water over the second weir. Determine discharge of 90° V-notch having 30 cm head of flow.

60

Measurement of Irrigation Water by Using Orifice

Aim

- To measure the irrigation water by using orifice

Orifices

An orifice is an opening in a bulk head with closed perimeter usually circular or rectangular cross-section through which water flows. For measurement of water, orifices are fabricated by making an accurate cut of proper size and shape in a mild steel sheet, aluminum plate etc. Proper machining is done to have a sharp edge through which water flows (Fig.60.1).

Free discharge

If the downstream water level is lower than the bottom of the orifice opening, the water jet discharges freely into the air and then it is said to be free discharge. The depth of water on the upstream side of the orifice that causes the flow is called the head. The orifice flow formulae can be applied only when the orifice flows fully.

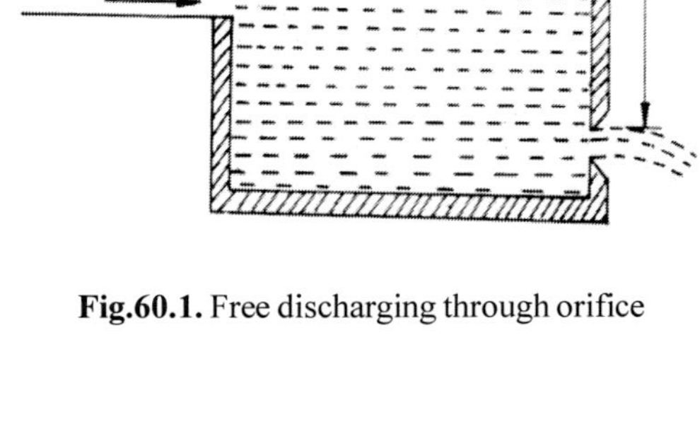

Fig.60.1. Free discharging through orifice

Discharging through orifice

$$Q = C_d \times 10^3 \times a\sqrt{2gh}$$

Where,

Q = discharge, lit/s

C_d = Coefficient of discharge, usually taken as 0.62 for sharp edged orifices

a = Cross-sectional area of orifice, cm^2

g = acceleration due to gravity, cm/s^2

H = head of water causing the flow, cm

Submerged orifice

If the orifice discharges under a condition of submergence, it is called submerged orifice. The submerged orifice generally has a rectangular shape. In sub-merged orifice, the outlet side is fully sub-merged under the water and it discharges a jet of water into the water of the same kind (Fig,60.2). It is also called totally drowned orifice.

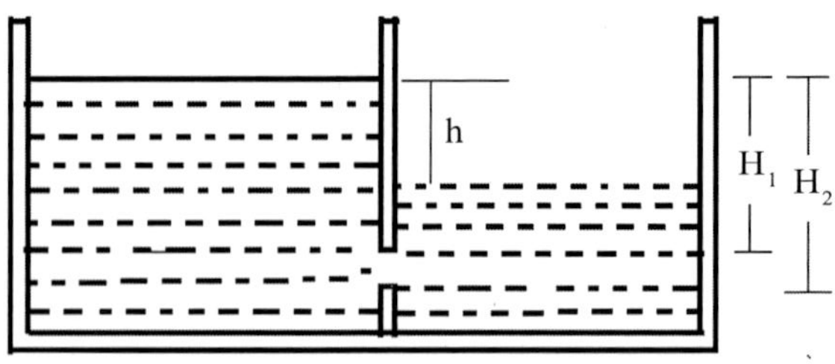

Fig.60.2. Submerged orifice

Discharge through fully sub-merged orifice:

$$Q = C_d \times b(H_2 - H_1) \times \sqrt{2gh}$$

Where,

H_1= height of water above the top of the orifice on the upstream side, cm

H_2= height of water above the bottom of the orifice, cm

h= difference in water level, cm

b= width of the orifice, cm

Example 1: A circular orifice has a diameter of 10 cm and the depth of water on the upstream side is 17 cm above the bottom of the orifice. If the coefficient of discharge is 0.61, calculate the discharge for free flow condition.

Solution: Data given: H = Head above the centre of the orifice = 17 - 5 = 12 cm

$$a = \frac{\pi}{4} \times d^2 = \frac{3.14}{4} \times 10^2$$

$= 78.54\ cm^2$

The discharge through an orifice can be calculated by the following formula:

$$Q = C_d \times 10^3 \times a\sqrt{2gh}$$

$$Q = 0.61 \times 10^3 \times 78.54\sqrt{2 \times 980.6 \times 12} = 7.35\ \text{lps}$$

Problem 2: Find the discharge through a fully sub-merged orifice of width 2 m if the difference of the water levels on the both sides of the orifice is 50 m. The height of water from top and bottom of the orifice are 2.5 m and 2.75 m respectively. Take Cd= 0.6.

Results

(i) Free discharging through orifice is __________

(ii) Discharge through fully sub-merged orifice is _________.

61

Measurement of Irrigation Water by Parshall Flume

Aim

- To measure the irrigation water by using parshall flume

Construction of Parshall flume

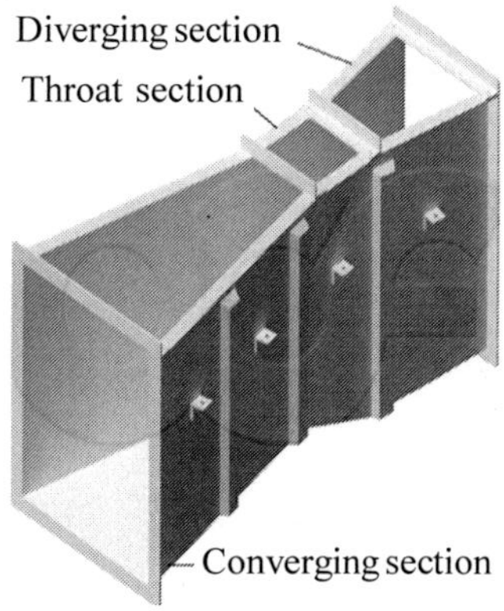

Fig.61.1. Parshall flume

Parshall flume is a water measuring device that can be used to measure water flowing in an open channel and intended primarily for use in irrigation. The flume consists of a converging inlet section, a throat section with straight parallel sides, and a diverging outlet section (Fig.61.1). The floor of the converging section is level with walls which converge in the direction of flow, towards the throat. The throat section constitutes the middle portion of the Parshall flume. The floor of the throat section is dips downward in the direction of flow, but the walls are parallel. The throat width specifies the dimensions of the parshall flume. The diverging outlet section of the flume has a floor which rises in the direction of flow. The walls become further separated in the direction of flow (diverge).The parshall flume can be constructed by G.I. sheet, aluminum sheet, M.S. sheet, brick masonry, reinforced concrete or timber. Flumes made of G.I. or aluminum sheets are easily portable. Brick masonry and concrete flumes are permanently constructed in the channels. Stilling wells should be constructed at the side to take measurements.

Terminologies

i) Flume crest

The flume crest is the edge of the floor at the junction of the converging inlet section and the throat section. Two types of flow conditions can exist in a Parshall flume, free flow and submerged flow.

ii) Free flow

Free flow exists in a Parshall flume when a change in the water surface elevation downstream from the flume does not cause a change in water surface elevation upstream from the flume. When free flow exists, only the depth of flow at Ha need be measured to determine the flow rate or discharge.

iii) Submerged flow

Submerged flow exists in a Parshall flume when a change in the water surface elevation downstream from the flume causes a change in water surface elevation upstream from the flume. When submerged flow exists, the depths of flow at both H_a and H_b must be measured to determine the flow rate or discharge. H_a is the depth of flow measured in the converging inlet section at a point two-thirds the length of the entrance section upstream from the flume crest. H_b is the depth of flow measured at a specified point in the throat. The value of H_b is the depth of flow above the elevation of the flume crest (Fig.61.2).

iv) Head loss

Head loss is the loss of energy of the flow between two points. The head loss can be expressed as the difference between H_a and H_b. This is only assumption usually made for Parshall flumes.

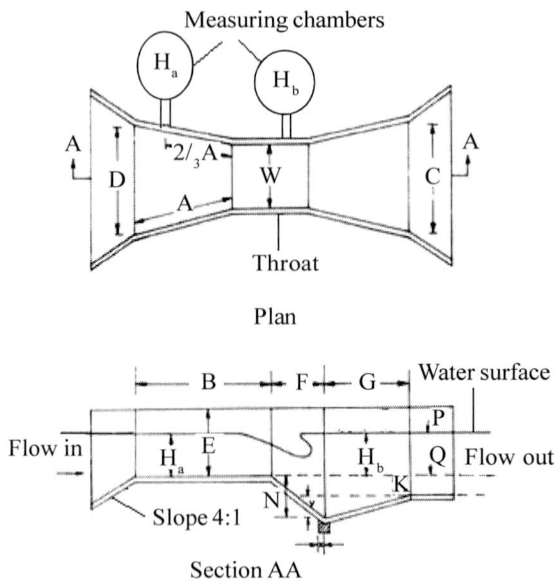

Fig.61.2. Schematic diagram of Parshall flume

Dimensions of Parshall flumes

The size of a Parshall flume is denoted by its throat-width and indicated as "W". It can be identified by both a section that has a contraction in width of the flume as well as a dip in the elevation of bottom of the flume. The height change is the difference in height between the upstream and downstream water heights. The converging floor section must be level in all directions. The floor of the throat is inclined downward with a slope of 23 cm vertically, to 61 cm horizontally. The floor of the outlet or diverging section has a slope upward of 15 cm vertically to 91 cm horizontally, with the downstream end of the flume 7.6 cm lower than the crest. These dimensions are for flumes having a throat width of 30 cm or more.

Discharge measurement

The discharge is measured by allowing the stream of water to flow through a converged section of the flume with a depressed bottom. The head loss, which is very small due to this flow condition is measured accurately to obtain the discharge.

Discharge through the flume is called "free flow" when the water elevation downstream is not high enough to affect flow conditions upstream. However, when the water elevation downstream is high enough to affect conditions upstream, submerged flow occurs.

Free flow is the condition under which the rate of discharge depends only upon the depth of flow, H_a in the converging section. The free flow discharge relationship remains valid for a relatively high degree of submergence, thus allowing a wide range of downstream flow conditions.

The submergence, is the ratio of the downstream head (H_b) to the upstream head (H_a), which is often expressed as percentage. The point at which the transition from free flow to submerged flow occurs can be observed in a Parshall flume. When downstream flow conditions cause the water surface in the flume to be raised until a ripple or wave is formed at or near the downstream end of the throat, the flow is near the transition submergence.

The equations used to determine the flow rate vary depending on the throat width of the flume.

For 3 inch throat width, $Q = 0.992 \times (H_a)^{1.547}$

For 6 inch throat width, $Q = 2.06 \times (H_a)^{1.58}$

For 9 inch throat width, $Q = 3.07 \times (H_a)^{1.53}$

For 1 foot throat width, $Q = 3.95 \times (H_a)^{1.55}$

For 2 foot throat width, $Q = 4W \times H_a^{1.522} \times W^{0.026}$

For 10-20 foot throat width, $Q = (3.6875W + 2.5)H_a^{1.6}$

Where,

Q = flow rate, cubic feet per second

H_a = height, ft

W = throat width, ft

Result

62

Measurement of Irrigation Water Flowing Through A Pipe by Venturimeter

Aim

- To measure the irrigation water flowing through a pipe by using Venturimeter

Theory

A venturimeter is a device used to measure the rate of flow of a fluid through a pipe and is often fixed permanently at different sections of the pipeline to know the discharge there. Venturimeter consists of three parts: (i) a short converging part (ii) a throat (iii) a diverging part (Fig.62.1). In the venturimeter, the fluid is accelerated through a converging cone of angle 15-20° and the pressure difference between the upstream side of the cone and the throat is measured, and provides the signal for the rate of flow. Due to the constriction there is an increase in the flow velocity and hence an increase in the kinetic energy. The fluid slows down in a cone with smaller angle (5-7°) where most of the kinetic energy is converted back to pressure energy. Because of the cone and the gradual reduction in the area there is no "vena contracta". The flow area is at minimum at the throat.

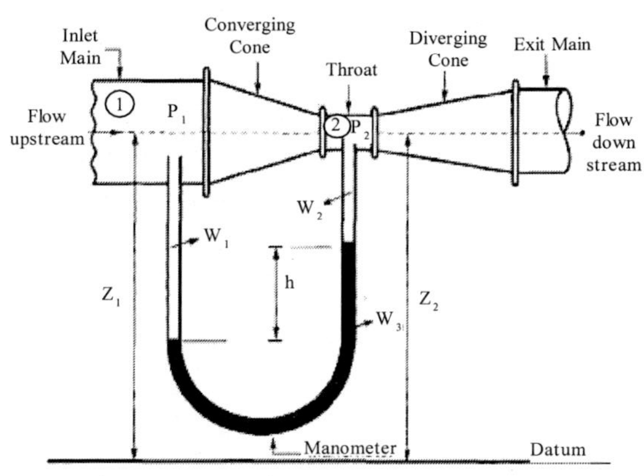

Fig.62.1. Schematic diagram of the venturimeter

Principle of operation

The venturimeter works on the Bernoulli's principle. From the Bernoulli's principle the increase in kinetic energy gives rise to a reduction in pressure. The rate of discharge from the constriction can be calculated knowing the pressure reduction across the constriction, area of cross section, density of fluid and the coefficient of discharge. The coefficient of discharge is the ratio of actual flow to the calculated flow, and it takes into account the stream contraction and frictional effects. The discharge in pipe flow can be measured by applying the Bernoulli's equation.

Bernoulli's equation

The application of the principle of conservation of energy to frictionless laminar flow leads to a very useful relation between pressure and flow speed in a fluid. This relation is called Bernoulli's equation, named after Daniel Bernoulli (1700-1782), who published his studies on fluid motion in his book *Hydrodynamica* (1738).

Consider an incompressible fluid flowing through a pipe that has a varying diameter and height, as shown in Fig.10.9. Also assume the density of fluid at the two points is the same, no friction, pipe is in horizontal, no pumping thus volume must be equal at two points. The points 1 and 2 in the figure denote two locations along the pipe and illustrate the relationships between the areas of the cross sections (A), the speed of flow (V), the height from ground (Z), and the pressure P at each point.

$$P_1 + \frac{1}{2}\rho V_1^2 + \rho g Z_1 = P_2 + \frac{1}{2}\rho V_2^2 + \rho g Z_2 \qquad (1)$$

This relation states that the mechanical energy of any part of the fluid changes as a result of the work done by the fluid external to that part, due to varying pressure along the way. Since the two points were chosen arbitrarily, the Bernoulli's equation can be written more generally as a conservation principle along the flow. For an incompressible, frictionless fluid, the combination of pressure and the sum of kinetic and potential energy densities is constant not only over time, but also along a streamline:

$$P_1 + \frac{1}{2}\rho V^2 + \rho g Z = \text{cons tan t} \qquad (2)$$

Procedure

The manometer in part (a) is connected to two tubes that are small enough not to appreciably disturb the flow. The tube facing the incoming fluid creates a dead spot having velocity (V_1) in front of it, while fluid passing the other tube has velocity V_2. Since, $Z_1 = Z_2$. This means that Bernoulli's principle as stated in:

$$P_1 + \frac{1}{2}\rho V_1^2 = P_2 + \frac{1}{2}\rho V_2^2 \tag{3}$$

$$P_1 - P_2 = \frac{1}{2}\rho\left(V_2^2 - V_1^2\right) \tag{4}$$

By rearrange the equation (4), we get

$$V_2^2 = \frac{2\left(P_1 - P_2\right)}{\rho} + V_1^2 \tag{5}$$

According to discharge, $Q = AV$ (6)

Therefore, discharge at point 1 (pipe) and point 2 (throat):

$$A_1 V_1 = A_2 V_2 \tag{7}$$

$$\frac{\pi d_1^2}{4} \times V_1 = \frac{\pi d_2^2}{4} \times V_2 \tag{8}$$

$$V_2 = \frac{d_1^2}{d_2^2} \times V_1 \tag{9}$$

$$V_2^2 = \left(\frac{d_1}{d_2}\right)^4 \times V_1^2 \tag{10}$$

By equating and simplifying the Eqs. (5) and (10), we can get

$$V_1 = \sqrt{\frac{2\left(P_1 - P_2\right)}{\rho\left(1 - D^4\right)}} \tag{11}$$

$$Q = A_2 \times \sqrt{\frac{2\left(P_1 - P_2\right)}{\rho\left(1 - D^4\right)}} \tag{12}$$

The coefficient of discharge is introduced in the above equation (12) an account for the friction losses depends upon the type of flow, type of fluid and dimensions of venture tube and pipe and the final equation becomes:

$$Q_a = C_d \times A_2 \times \sqrt{\frac{2\left(P_1 - P_2\right)}{\rho\left(1 - D^4\right)}} \tag{13}$$

Where,

$$D = \frac{d_1}{d_2}$$

A_2 = cross sectional area of the throat

d_1 = diameter at point 1 (pipe)

d_2 = diameter at point 2 (throat)

ρ = density of water

P_1 = water pressure at point 1

P_2 = water pressure at point 2

C_d = coefficient of discharge

Calculation of head difference

The head difference in terms of the liquid flowing through the venturi from the head difference observed in the manometric liquid can be calculated as given below. If the tube facing the incoming fluid creates a dead spot having zero velocity ($V_1 = 0$) in front of it, while fluid passing the other tube has velocity V_2. Since, $Z_1 = Z_2$.

$$P_1 = P_2 + \frac{1}{2}\rho V_2^{\,2} \quad (14)$$

Thus pressure P_2 over the second opening is reduced by $\frac{1}{2}\rho V_2^{\,2}$

The fluid in the manometer rises by 'h' on the side connected to the second opening is:

$$h \alpha \frac{1}{2}\rho V_2^{\,2} \quad (15)$$

By resolving the equation, we can get, $V_2 \propto \sqrt{h}$

1. When the manometric liquid is heavier than the liquid flowing through the pipe:

$$h = \chi\left(\frac{S_{pl}}{S_{ml}} - 1\right) \quad (16)$$

2. When the manometric liquid is lighter than the liquid flowing in the pipe:

$$h = \chi\left(1 - \frac{S_{pl}}{S_{ml}}\right) \quad (17)$$

Where,

h = head difference in terms of the liquid flowing in the pipe

x = head difference in the manometer

S_{pl} = specific gravity of the liquid flowing in the pipe

S_{ml} = specific gravity of the manometric liquid

Example

Determine the flow rate of irrigation water by using a horizontal venturimeter with inlet and throat diameters 30 cm and 15 cm respectively. The reading of the differential manometer connected to the inlet and the throat is 12 cm of mercury. Assume C_d is 0.86, specific gravity of the water flowing in the pipe and manometric liquid are 1.0 and 0.85 respectively.

Given: $x = 12$ cm of mercury; $d_1 = 30$ cm; $d_2 = 15$ cm; $S_{pl} = 1$; $S_{ml} = 0.85$

Area of throat,

$$a_2 = \frac{\pi {d_2}^2}{4} = \frac{\pi \times 15^2}{4} = 176.625\,\text{cm}^2$$

$$D = \frac{{d_1}^2}{{d_2}^2} = \frac{30}{15} = 2$$

$$Q_a = C_d \times A_2 \times \sqrt{\frac{2(P_1 - P_2)}{\rho(1 - D^4)}}$$

$$\frac{2(P_1 - P_2)}{\rho} = 2gh$$

Manometer liquid is heavier than water when mercury is liquid. So, Eq. (16) is used

$$h = \chi\left(\frac{S_{pl}}{S_{ml}} - 1\right)$$

$$h = 12\left(\frac{1}{0.85} - 1\right) = 2.1176\,\text{cm of water}$$

$$2gh = 2 \times 981 \times 2.1176 = 4154.824$$

$$Q = 0.86 \times 176.625 \times \sqrt{\frac{4154.824}{(1 - 2^4)}} = 2.528 \times 10^{-3}\,\text{m}^3/\text{s}$$

Result

i) Discharge through a pipe by using Venturimeter is ____________

ii) The head difference in terms of the liquid flowing through the venturimeter is _____

63

Design of Grassed Water Ways or Open Channel

Aim

- To design the grassed water ways or open channel

Geometric characteristics of different waterways

- The shape of open channel is dependent upon the field conditions and types of the construction equipments used.
- The common shapes of open channels generally used are (i) Parabolic or semi circular (ii) Triangular (iii) Rectangular and (iv) Trapezoidal section.
- The parabolic shape approximates a natural channel. The reason is that in natural course of channel flow, the trapezoidal and triangular sections tend to become in parabolic shape due to deposition of sediments and bank erosion.

i) Parabolic or semi circular

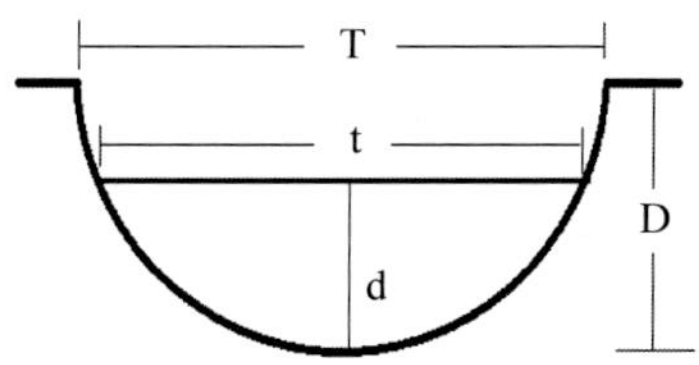

Fig.63.1. Geometric characteristics of parabolic section

$$\text{Area}\,(A) = \frac{2}{3} \times t \times d$$

$$\text{Wetted perimeter}\,(P) = t + \frac{8d^2}{3t}$$

$$\text{Hydraulic radius}\,(R) = \frac{t^2 \times d}{1.5t^2 + 4d^2}$$

$$\text{Top width of water level}(t) = \frac{A}{0.67 \times d}$$

$$\text{Top width of the channel}\,(T) = t \times \left(\frac{D}{d}\right)^2$$

ii) Triangular section

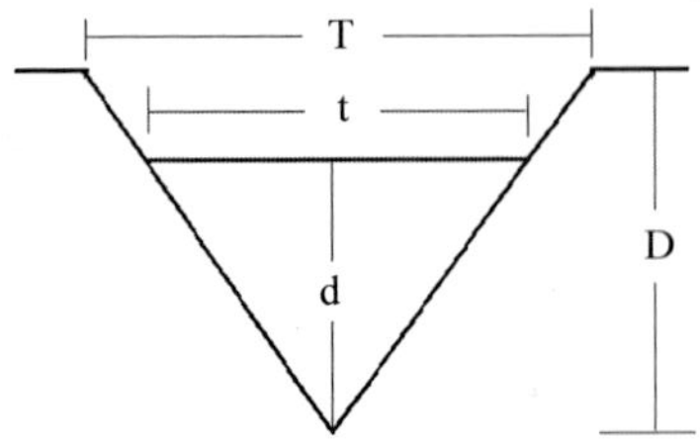

Fig.63.2. Geometric characteristics of triangular section

Area of triangular section (A) = $z \times d^2$

$$\text{Wetted perimeter (P)} = 2d\sqrt{z^2+1}$$

$$\text{Hydraulic radius (R)} = \frac{A}{P} = \frac{zd}{2\sqrt{z^2+1}}$$

Top width up to flow (t) = $2 \times d \times z$

$$Top\ width\ (T) = \frac{D}{d} \times t$$

iii) Rectangular section

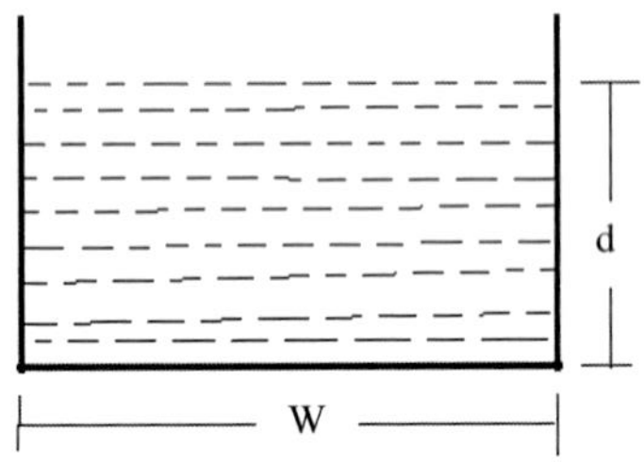

Fig.63.3. Geometric characteristics of rectangular section

Area, A = W × d

Wetted perimeter, P = W + 2d

Hydraulic radius, R = A/P

In rectangular section, the best hydraulic section exist when depth is equal to half width.

$$d = \frac{W}{2}$$

iv) Trapezoidal section

Among these channels trapezoidal section is commonly used, as these are easy to construct and maintain. The geometric characteristics of the trapezoidal waterway are shown in Fig.63.4. These characteristics formulae are useful for design purposes.

Let,

T = top width of the channel, m

t = width of surface when the water is at depth d, m

D = total depth of channel after free board is added, m

y or d = depth of flow in channel, m

b = bottom width of channel, m

L = wetted sides of channel, m

f = free board

θ = angle between the sloping sides and the horizontal (ie, angle of repose)

Area of cross-section: It refers to the area of the wetted section of the channel.

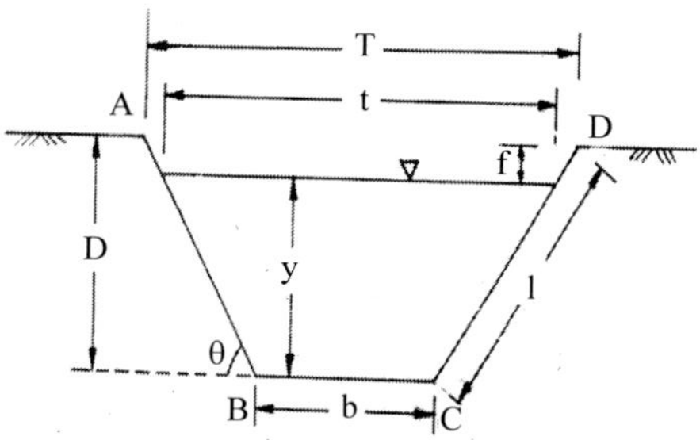

Fig.63.4. Geometric characteristics of trapezoidal waterway

$$A = \frac{(b+t)}{2} \times d$$

Wetted perimeter: It is the sum of the lengths of sides and bottom that are in contact with water

P = b + 2L

Hydraulic radius: It is the ratio between cross-sectional area of the flow and the wetted perimeter. Hydraulic radius increases with decrease in wetted perimeter for a given cross-sectional area.

$$R = \frac{A}{P}$$

Hydraulic slope: Hydraulic slope (S) of a channel is expressed as the ratio of the vertical drop (h) to the length (l) of the channel.

$$S = \frac{h}{l}$$

Free board

It is the vertical distance between the top of the channel wall and maximum possible water level in the channel. It is required as a safety measure against overtopping due to wave action or other reasons.

Angle of repose (θ)

- If mass of loose dry soil is allowed to fall freely to the ground, the soil heap is formed and it will attain a conical shape.
- The maximum angle of the slopping sides of the cone measured from the horizontal is called the angle of repose.
- Channel sides will be stable if its slope does not exceed the angle of repose of the soil in which it is made. Clay soils have steeper angle of repose which in sandy soils they are flatter.

Velocity of flow

The design of channels is based on the mean velocity of flow which is obtained by standard formulae.

i) Chezy's formula

$$v = c \times \sqrt{R \times S}$$

ii) Manning's formula

$$V = \frac{R^{2/3} \times S^{1/2}}{n}$$

Where,

V = Velocity of flow, m/s

R = Hydraulic Radius, m

S = Hydraulic slope, dimensionless

n = Manning's roughness coefficient

C = Chezy's constant

Discharge: The discharge can be calculated using the following formula:

$Q = A \times V$

Where,

Q = discharge capacity, m^3/s

A = cross sectional area of the wetted section of the channel, m^2

V = mean velocity of the flow in channel, m/s

Most economic channel section

- An economic channel section is one which carries the maximum discharge for a given cost of construction. Maximum discharge is obtained when the velocity is maximum.
- In a rectangular section, half a square (depth equal to half width) is the best hydraulic section.

$$d = \frac{w}{2}$$

- In trapezoidal section, the best hydraulic section is obtained from the following relationship:

$$b = 2d \tan\left(\frac{\theta}{2}\right)$$

Where,

b = bed width, (m)

d = depth of flow, (m)

θ = angle between the side and the horizontal.

Stepwise procedure for design of trapezoidal channel section

The design of open channel includes: (i) Determination of channel cross-section for a given discharge (ii) Determination of velocity of flow within permissible limit for the given soil condition. The open channel may be rectangular, trapezoidal, semi-circular/parabolic shape. Among these channels trapezoidal section is commonly used, as these are easy to construct and maintain.

i) Width of the channel when water is at depth 'd'

$$t = b + (2 \times z \times d)$$

$$z = \frac{\text{Horizontal}}{\text{Vertical}}$$

ii) Length of slopping side: $L = \sqrt{(zd)^2 + d^2} = d\sqrt{z^2 + 1}$

iii) Wetted perimeter: $P = b + 2d\sqrt{z^2 + 1}$ or $P = b + 2L$

iv) Area of channel cross-section:

$$A = \frac{(b+t)}{2} \times d$$

(or)

$$A = (b + z\,d)\,d$$

v) Hydraulic radius:

vi) Velocity of flow (using one of the following formulae):

a) Chezy's formula: $V = C \times \sqrt{R \times S}$

b) Manning's formula: $V = \frac{R^{2/3} \times S^{1/2}}{n}$

vii) Discharge, $Q = A \times V$ (m^3/s)

Problem 1: A trapezoidal earthen channel is constructed with a bed slope of 3 %. The bottom width is 2 m, depth of flow is 1 m and the side slope of the channel is 2:1. Calculate the velocity of flow and discharge using (i) Chezy's formula, Chezy's constant 'c' = 60 (ii) Manning's formula, Manning's roughness coefficient is 0.04.

Solution: $b = 2$ m, $d = 1$ m, $S = 3$ % $z = H/V = 2/1 = 2$

Length of slopping side, $L = d\sqrt{z^2 + 1} = 1\sqrt{2^2 + 1} = 2.23\text{m}$

Wetted perimeter, $P = b + 2L = 2 + (2 \times 2.236) = 6.472\text{m}$

Cross-sectional area, $A = (b + zd)\, d = (2 + 2 \times 1) \times 1 = 4\ \text{m}^2$

Hydraulic radius: $R = \dfrac{4}{6.47} = 0.62\text{m}$

Using Chezy's formula:

Velocity of flow, $V = C\sqrt{RS} = 60 \times \sqrt{0.618 \times 0.03} = 8.17\text{m/s}$

Discharge, $Q = AV = 4 \times 8.1699 = 32.679\ \text{m}^3/\text{s}$

Using Manning's formula:

Discharge, $Q = A \times V = 4 \times 3.142 = 12.568\ \text{m}^3/\text{s}$

Problem 2: A trapezoidal earthen channel is constructed with a bed slope of 1 in 3500. The bottom width is 2 m, depth of flow is 80 cm and the side slope of the channel is 2:1. Calculate the velocity of flow and discharge using (i) Chezy's formula, Chezy's constant 'c' = 60

Solution: $b = 2$ m, $d = 0.8$ m, $S = 1/3500$, $z = H/V = 2/1 = 2$

Length of slopping side, $L = d\sqrt{z^2 + 1} = 0.8 \times \sqrt{2^2 + 1} = 1.792\text{m}$

Wetted perimeter, $P = b + 2L = 2 + (2 \times 1.792) = 5.58$ m

Cross-sectional area, $A = (b + zd)\, d = (2 + 2 \times 0.8) \times 0.8 = 2.88\ \text{m}^2$

Hydraulic radius, $R = \dfrac{A}{P} = \dfrac{2.88}{5.58} = 0.516\text{m}$

Using Chezy's formula:

Velocity of flow, $V = C\sqrt{RS} = 60 \times \sqrt{0.516 \times \dfrac{1}{3500}} = 0.729\text{m/s}$

Discharge, $Q = AV = 2.88 \times 0.729 = 2.098\ \text{m}^3/\text{s}$

Problem 3: What would be the most economical cross-section for a rectangular channel? If gradient is 0.5 % chezy's C = 12 and depth of water may be taken as 1 m. calculate the velocity and discharge of the channel.

Solution: $d = 1$ m, $S = 0.5$ % $C = 12$

Condition for most economic rectangular section, $d = b/2,\ b = 2d$

Bottom width, $b = 2 \times 1 = 2$ m

Cross-sectional area, $A = b \times d = 2 \times 1 = 2\ m^2$

Wetted perimeter, $P = b + 2d = 2 + (2 + 1) = 4$ m

Hydraulic radius, $R = A/P = 2/4 = 0.5$ m

Velocity of flow, $V = C\sqrt{RS} = 12 \times \sqrt{0.5 \times 0.005} = 0.6\,m/s$

Discharge, $Q = AV = 2 \times 0.6 = 1.2\ m^3/s$

Problem 4: Design a most economic trapezoidal section of a channel to carry a discharge of 10 m^3/s in a bed slope of 1 in 3000. for stability, the side slope is 1.5:1. the value of Manning's n is 0.025.

Solution: $z = 1.5\ H : 1\ V = 1.5$, $Q = 1.5\ m^3/s$, $S = 1/3000$,

Condition for most economic trapezoidal section,

$$b = 2d\tan\left(\frac{\theta}{2}\right)\text{...............}(1)$$

$$\tan\theta = \frac{1}{1.5} = 0.667$$

$$\theta = \tan^{-1}(0.667) = 37.433;\ \frac{\theta}{2} = 18.716$$

$$\tan\left(\frac{\theta}{2}\right) = \tan(18.716) = 0.31$$

Substitute the value of tan (θ/2) in equation (1), we get

$$b = 0.6055\ d\text{.................}(2)$$

$$L = \sqrt{d^2 + (1.5d)^2}\text{ ....}(3)$$

Simplify this equation, we get

$$L = 1.803\ d\text{...............}(4)$$

Wetted perimeter, $P = b + 2L$...................(5)

Substitute the value of b and L in the equation (5), we get

$$P = 4.211\ d\text{..................}(6)$$

Cross-sectional area, $A = (b + zd)\,d$..................(7)

Substitute the value of b and z in the equation (7), we get

$$A = 2.1055\ d^2 \ldots\ldots\ldots\ldots\ldots\ldots (8)$$

Hydraulic radius, $R = \frac{A}{P} \ldots\ldots\ldots\ldots\ldots\ldots (9)$

Substitute the value of A and P in the equation (9), we get

$$R = 0.5\ d \ldots\ldots\ldots\ldots\ldots\ldots (10)$$

Discharge, $Q = A\ V$

$$Q = A \times \frac{R^{2/3} \times S^{1/2}}{n} \ldots\ldots\ldots\ldots\ldots\ldots (11)$$

Rearrange this equation,

$$AR^{2/3} = \frac{Qn}{S^{1/2}} \ldots\ldots\ldots\ldots\ldots\ldots (12)$$

Substitute the value of A, R, Q, n and S in the equation (12), we get

$$(2.1055d^2) \times (0.5d)^{2/3} = \frac{10 \times 0.025}{\sqrt{\frac{1}{3000}}}$$

$$1.326 \times d^{8/3} = 13.698$$

$$d = (10.327)^{3/8} = 2.4m$$

Substitute the value of 'd' in the equation (2), we get

$b = 0.6055 \times 2.4 = 1.4$ m

Add freeboard of 10 % of depth of flow or approximately 25 cm

Results

Total depth of the channel (D) = 2.4 + 0.25 = 2.65 m

Bottom width of channel (b) = 1.4 m

Depth of flow (d) = 2.4 m

Length of side slope (L) = 1.803 d = 1.803 x 2.4 = 4.327 m

Width of the channel (t) when water is at depth d

$t = b + 2zd = (1.4 + 2 \times 1.5 \times 2.4) = 8.6$ m

Exercise

1. Design an economical section of a trapezoidal channel to carry a discharge of 8.4 m^3/s. the bed slope is 1 in 4000 and the side slope is 2:1. The Manning's n is 0.015.
2. Assume an earth channel on a grade of 15 per cent, depth of water 40 cm, bottom width 30 cm and side slope 1.5 to 1. calculate the velocity of flow and carrying capacity of the channel. Manning's n is 0.025.
3. Determine the bottom width, depth of flow and mean velocity of flow in a channel with 2:1 side slope for a flow of 1.5 m^3/s. Assume S = 0.2 and Manning's n = 0.04
4. Design an earth channel to carry a flow of 55 l/s in loamy sand soil. The land slope favours a channels bed slope of 1 m per 1000 m. Manning n = 0.025

64

Design of Contour Bunding System

Aim

- To design the contour bunds

Design parameters

The design of contour bund includes determination of following parameters:

- Choice of bund
- Spacing of bunds
- Size of bund
- Alignment of bund

1. Choice of bund

The choice of bund should be selected weather the contour bund or graded bund depends on the annual rainfall, soil condition and types of outlets used.

2. Spacing between the bunds

The vertical interval between two successive bunds is said to be spacing of bunds. Horizontal distance between the bunds also decides the spacing. The basic principles which decide the spacing between the bunds are the phreatic zone below the upper bund should coincide the saturation zone of the immediate lower bund .

i) Vertical interval

The following relationship was developed by C.E Ramser for vertical interval of contour bunds.

$$V.I = \frac{S}{a} + b \text{.................} (1)$$

Where,

V.I = vertical interval between two successive bunds, m

S = land slope, percent

a, b = constants depends on the soil and rainfall characteristics of the area

When the formula is used for soils with high infiltration rate and good conservation practices, 25% extra spacing can be used. In soils of low infiltration capacity and unfavorable conservation measures, the spacing should be reduced by 15%.

a) For the areas having heavy rainfall

$$V.I = \frac{0.3S}{3} + 0.6 \text{...............} (2)$$

b) For the areas having low rainfall

$$V.I = \frac{0.3S}{3} + 0.6 \text{..............} (3)$$

ii) Horizontal interval

The following relationship is used to determine the horizontal spacing of the bund.

$$H.I = \frac{V.I}{S} \text{................} (4)$$

3. Size of bund

The size of bund includes height, bottom width, top width and side slopes. Generally, the shape of the bund is trapezoidal and therefore, cross-sectional area is

$$\text{Area} = (\text{bottom width} + \text{top width}) \times \frac{\text{height}}{2}$$

i) Determination of height of bund

- Once the height of bund is determined, the other dimensions such as top width and bottom width may easily be obtained.
- The height of bund mainly depends on the spacing of bunds, depth of water to be stored, design depth of flow over waste weir, land slope and maximum expected rainfall intensity of the area.
- The rate of infiltration influences the vertical interval and depth of water to be impounded.

Height of bund is determined on the basis of amount of water to be intercepted by the bund (Fig.64.1).

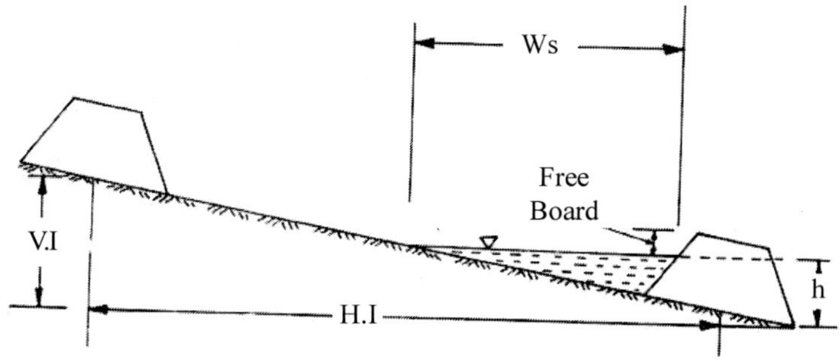

Fig.64.1. Height of contour bund

Let,

H.I = horizontal interval between the successive bunds, m

V.I = vertical interval between two bunds, m

W_s = water spread length behind the bund, m

H = depth of water impounding near the bund, m

R = 24-hours excess rainfall to be stored, cm

R_v = volume of runoff, m^3

S_v = storage volume of bund of one metre wide strip, m^3

S = land slope, percent

Horizontal interval

The horizontal interval is related with the vertical interval by the following formula:

$$H.I = \frac{V.I}{S} \times 100 \ldots\ldots\ldots\ldots\ldots\ldots (5)$$

Volume of runoff

Referring to Fig.64.1, the volume of runoff from 1 m width of the land between two bunds is given by

$$R_v = \frac{R}{100} \times H.I \ldots\ldots\ldots\ldots\ldots (6)$$

Volume of excess rainfall (R) can be calculated from the hydrologic soil cover complex method.

Storage volume of water

The storage volume of water behind the bund is given by following relationship. Neglect minor effects due to the side slopes of the bund.

$$S_v = \frac{W_s \times h}{2} \quad \text{................(7)}$$

Equating R_v and S_v from the above relationships, we have

$$\frac{W_S \times h}{2} = \frac{R \times HI}{100}$$

$$h = \frac{R}{50} \times \frac{HI}{W_s} \quad \text{.................(8)}$$

From the similar triangles of Fig.64.1, we have

$$\frac{W_s}{HI} = \frac{h}{VI} \Rightarrow W_s = \frac{HI \times h}{VI} \quad \text{...........(9)}$$

Substituting the value of W_s in equation (8) and simplify, we get,

$$h = \sqrt{\frac{R \times VI}{50}} \quad \text{.................(10)}$$

Equation (10) gives the theoretical depth of water that will be impounded immediately behind the bund. To get total height of bund, a freeboard of 20-25% of the theoretical depth should be added. Generally, a depth of flow of 30 cm is considered above the weir.

ii) Cross-section of bund

Cross-section such as top width and bottom width should be decided after fixing the height. The side slopes are decided from angle of repose of the soil. Recommended typical cross-sections of the contour bunds by Gadkary (1954) are presented in Table 64.1.

Table 61.1. Typical cross-sections of contour bund

Soil type	Top width, m	Bottom width, m	Height m	Side slope	Area m^2
Full murum or soil layer upto 7.5 cm	0.45	2.00	0.75	1 : 1	0.919
Soil layer from 7.5 to 22.5 cm	0.45	2.60	0.83	1.25 : 1	1.265
Full soil or soil layer from 22.5 to 45 cm	0.53	3.12	0.83	1.5 : 1	1.486
Full soil 45 to 90 cm	0.60	4.50	0.90	2 : 1	2.295

iii) Side slope of the bund

Recommended values of side slopes for different types of soils are listed in Table 64.2. A suitable size of bund is selected based on the soil characteristics and rainfall intensity.

Table 64.2. Recommended side slopes for different soil types

S.No	Soil types	Side slope (H:V)
1.	Red, gravel, light red, loam, black loam, white gravel	1.5 : 1
2.	Light sandy loam, clay, black cotton soil, soft decomposed rock	2 : 1
3.	Sandy soil	2.5 : 1

4. Alignment of contour bunds

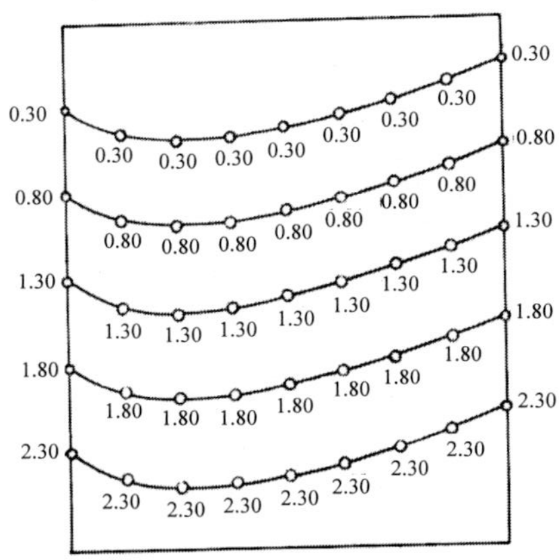

Fig.64.2. Alignment of contour bunds

- Level survey should be conducted by taking a suitable grid distance so that contours can drawn on the map at an interval of 30 cm.
- Average slope of the land is calculated by using the contour map. Vertical and horizontal intervals are calculated by knowing the land slope.
- Planning is started from the top of the watershed and contour bunds are located on the map as per the designed horizontal and vertical intervals.
- Long length of bunds should be avoided. The length of bund should not exceed 350 m in one direction of surface flow.

5. Outlets of contour bunds

Surplusing arrangement should be provided for safe disposal of excess runoff from the bunded area, both in contour and graded bund system. The surplusing arrangements are

i) clear over fall weir

ii) pipe outlet

i) Clear over fall weir

The clear overfall weir (Fig.64.3) is a permanent masonary structure. The crest height of weir controls the depth of water collected toward its upstream face. A crest height of 30 cm above the contour bund is recommended and head of flow is limited upto 30 cm.

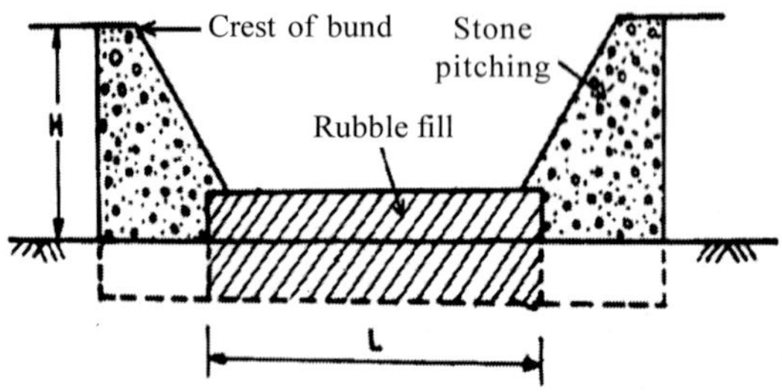

Fig.64.3. Clear over fall weir

Following formula is used for calculating the length of weir:

$$Q = CLH^{3/2} \qquad (11)$$

Where,

Q = peak runoff rate

L = length of the weir

H = depth of flow water over the crest

C = weir coefficient

ii) Pipe outlet

Pipe outlet weir consists of upstream (inlet) and downstream (outlet) stilling wells connected with a conduit pipe, made of reinforced cement concrete or cast iron of required diameter that buried under the bund with a downward slope (Fig.64.4). Generally, a 15 cm diameter pipe is suitable for 3.5 ha catchment and 30 cm diameter for 8 ha catchment.

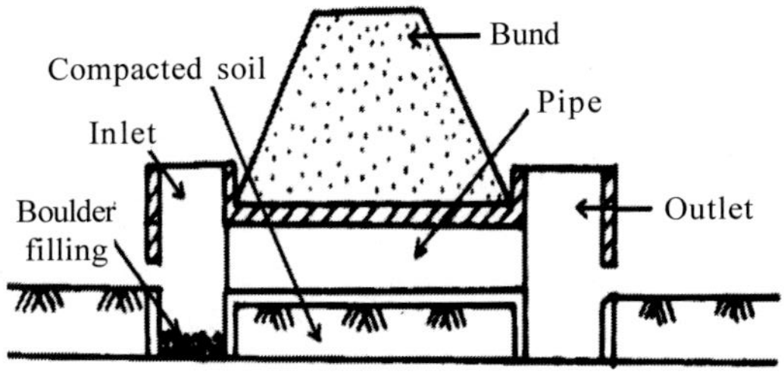

Fig.64.4. Pipe outlet

6. Earthwork computation and area lost for contour bunding

The earthwork includes main contour bund, side and lateral bunds. The length of side and lateral bunds may be 30% of the length of contour bund.

Assume, L = length of contour bund

$$HI = \frac{VI}{S} \times 100 \ldots\ldots\ldots\ldots(12)$$

Number of contour bunds (N) per hectare of land is given by

$$N = \frac{10^4}{L \times HI} \ldots\ldots\ldots\ldots(13)$$

Substitute Eqn. (12) in the Eqn. (13), we get.

$$N = \frac{100S}{L \times VI} \ldots\ldots\ldots\ldots(14)$$

Length of contour bund per hectare of land (assume, number of bund is one)

$$L = \frac{100S}{VI} \ldots\ldots\ldots\ldots(15)$$

Adding 30% length for side and lateral bunds, total length of bund per hectare

$$L = \frac{130S}{VI} \ldots\ldots\ldots\ldots(16)$$

Cross-sectional area of the bund

$$A = \frac{(b+T)}{2} \times h \ldots\ldots\ldots\ldots(17)$$

Volume of earthwork per hectare of bunding

$$V = \frac{130S}{VI} \times \frac{(b+T)}{2} \times h \ldots\ldots\ldots\ldots(18)$$

Area lost due to bunding per hectare = (total length of bund) x (base width of bund)

$$= \frac{130S}{VI} \times b \text{..........(19)}$$

Percentage area lost due to bunding

$$= \frac{130S \times b \times 100}{VI \times 10^4} = \frac{1.3Sb}{VI} \text{..........(20)}$$

Example 1: Calculate the cross-section of a contour bund used to store a 24-hour excess rainfall of 15 cm. The slope of land is 4.5%, annual rainfall is about 150 cm, the soil is murrum with a high infiltration rate and there is a low coverage of crops over the land.

Solution: Given data: R = 15 cm, S = 4.5%

$$V.I = \frac{0.3S}{3} + 0.6 = \frac{0.3 \times 4.5}{3} + 0.6 = 1.05\text{m}$$

$$h = \sqrt{\frac{R \times VI}{50}} = \sqrt{\frac{15 \times 1.05}{50}} = 0.561\text{m}$$

Assume, freeboard = 0.20 m and depth of flow over the waste weir is 0.30 m

Therefore, total height = 0.561 + 0.2 + 0.3 = 1.061 m

For murrum soil, a top width of 0.45 m, side slope of 1:1 are assumed.

Therefore, base width = t + 2zd = 0.45 + (2 × 1 × 1.061) = 2.572 m.

Example 2: Design a suitable contour bund in medium soil having an average slope of 2%. The maximum expected rainfall during the 10-year recurrence interval is 15 cm. The infiltration capacity of the soil of the area is such that 40% of the rainfall is absorbed in the field. The horizontal interval between bunds is 60 m. Assume side slope of the bund is 1.5:1 and slope of seepage line is 4:1.

Solution:

Maximum rainfall during the 10-year recurrence interval = 15 cm

Infiltration capacity = 40% of maximum rainfall

Horizontal interval (HI) = 60 m

Land slope = 2%

Maximum runoff to be stored in the bund (R) = (Rainfall – Infiltration)

= 15 – (15 × 0.40) = 9 cm

$$\text{Storage area required (A)} = \frac{R \times H.I}{100} = \frac{9 \times 60}{100} = 5.4 m^2$$

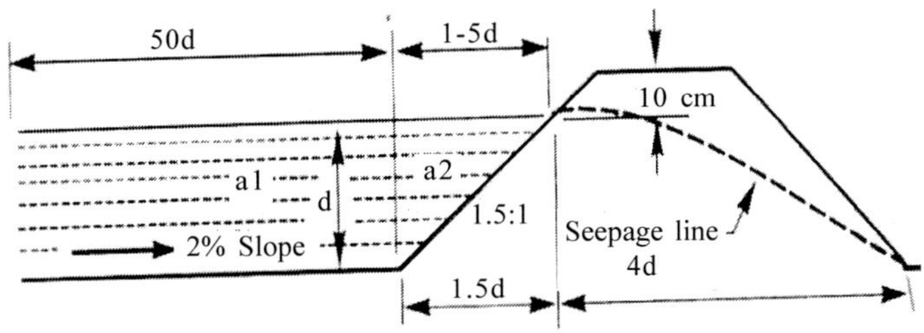

Fig. 64.5. Design of contour bund

Referring to Fig.64.5,

$$a_1 = \frac{1}{2} \times 1.5d \times d = 0.75d^2$$

$$a_2 = \frac{1}{2} \times 50d \times d = 25d^2$$

$$a_1 + a_2 = A$$

$$0.75d^2 + 25d^2 = 5.4$$

$$d = \sqrt{\frac{5.4}{25.75}} = 0.46m$$

(or)

$$V.I = \frac{S \times H.I}{100} = \frac{2 \times 60}{100} = 1.2m$$

$$h = \sqrt{\frac{R \times VI}{50}} = \sqrt{\frac{9 \times 1.2}{50}} = 0.465m$$

Total height of bund (freeboard) = theoretical height + 20% of theoretical height

= 0.465 + (0.20 × 0.465) = 0.558 m or 56 cm

Base width = 4d + 1.5d = (4 × 0.46) + (1.5 × 0.45) = 2.5 m

Top width = 2.5 – (1.5D +1.5D)

= 2.5 – (1.5 × 0.56 + 1.5 × 0.56) = 0.82 m

65

Design of Graded Bunds

Aim

- To design a graded bunds

The procedure in design of contour bund is followed while design a graded bund. A stability of 10-25%, should be given to contour bund. This is called graded bund.

Example: Design a graded bund length of 500 m with an average land slope of 3.5% in a sandy loam soil. Vertical distance between consecutive bund is 2.5 m. Rainfall intensity for the time of concentration and for the recurrence interval is 18.5 cm/hr. The runoff coefficient is 0.3. For stability, the grade provided for first 100 m is 0.1%, for next 200 m is 0.15% and for last 200 m is 0.18%. For the sandy loam soil, side slope of the bund and slope of seepage line are 1.5:1 and 5:1 respectively is assumed.

Solution: L = 500 m, S = 3.5%, VI = 2.5 m, I = 18.5 cm/hr, C = 0.3

Horizontal interval between two bunds

$$H.I = \frac{V.I \times 100}{S} = \frac{2.5 \times 100}{3.5} = 71.43\text{m}$$

Catchment area formed by 2 bunds = (500 × 71.43) /10000 = 3.57 ha

Design runoff rate,

$$Q = \frac{CIA}{36} = \frac{0.3 \times 18.5 \times 3.57}{36} = 0.55\text{m}^3/\text{s}$$

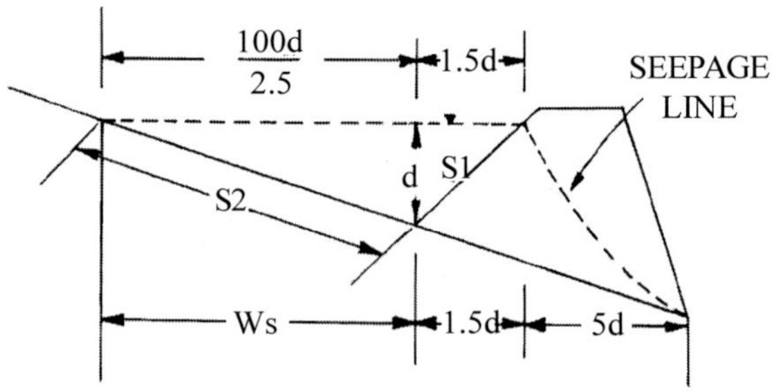

Fig.65.1. Design of graded bund

Referring to Fig.65.1,

Maximum depth of water impounded near the bund (d) = 30 cm

Width submerged along the upstream

$$W_s = \frac{100d}{S} = \frac{100 \times 0.3}{3.5} = 8.57m$$

Water cross-section,

$$A = a_2 + a_1 = \left(\frac{1}{2} \times W_s \times d\right) + \left(\frac{1}{2} \times 1.5d \times d\right)$$

$A = (0.5 \times 8.57 \times 0.3) + (0.5 \times 1.5 \times 0.3 \times 0.3) = 1.353\ m^2$

Wetted perimeter,

$$p = s1 + s2 = \sqrt{0.32 + 0.452} + \sqrt{8.572 + 0.32} = 9.1158m$$

Hydraulic radius,

$$R = \frac{a}{p} = \frac{1.353}{9.1158} = 0.1484m$$

velocity of water,

To calculate the velocity of water (velocity is non-erosive), Manning's equation is applied

$$V = \frac{1}{n} R^{2/3} S^{1/2} = \frac{1}{0.04} \times 0.1484^{2/3} \times 0.0018^{1/2} = 0297 m/s$$

Channel capacity, Q= AV = 1.353 × 0.2973 = 0.4022 m^3/s

The designed capacity is lower than required capacity.

Therefore the height of bund should be more than 30 cm. A freeboard of 20 cm is allowed.

Total height of bund (H) = 30 + 20 = 50 cm or 0.5 m

Base width for seepage line when the water depth is 30 cm

$= 30 \times 5$ cm = 150 cm (or) 1.5 m.

Therefore, required base width of the bund (b) = 1.5 + 0.5 = 2 m

Top width of bund (t) = b – (2zH) = 2 – (2 × 1.5 × 0.5) = 0.5 m.

66

Design of Bench Terracing System

Aim

- To design bench terrace

1. Components of bench terrace

The bench terrace essentially consists of the four components namely: riser, outlet channel, platform and shoulder bund (Fig.66.1).

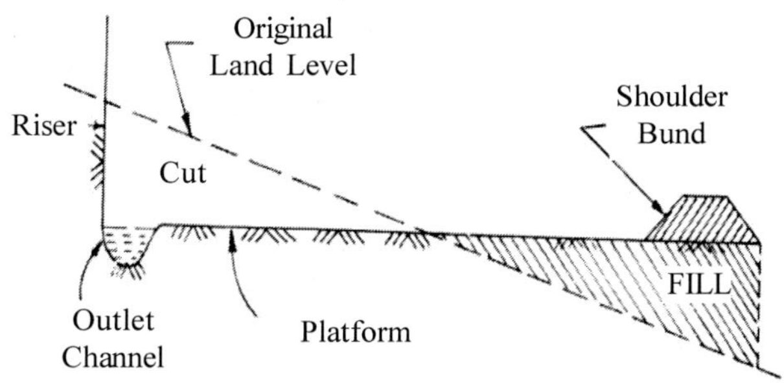

Fig.66.1. Components of bench terrace

i) Shoulder bund

Shoulder bunds are constructed at edge of the terrace bench to provide the stability to the outer edge of the terrace and also helps in retaining the surface runoff on the benches that is either absorbed by the soil or drained.

ii) Riser

It is vertical wall at upstream end of the strip of the land converted to bench terrace.

iii) Outlet

Outlet can be located either at the upstream or at the downstream end depending upon the rainfall and soil conditions.

iv) Platform

It is the level strip obtained by terracing and crop is grown in this zone.

2. Design of bench terrace

Design of bench terrace depends on the soil type, soil depth, average rainfall and land slope. The following parameters are involved for design of bench terrace

i) Type of bench terrace

ii) Spacing of bench terrace

iii) Width of bench terrace

iv) Terrace cross-section

v) Terrace grade and length

i) Type of bench terrace

Based on the annual average rainfall and permeability of the soil, the type of bench terrace to be used is selected. Types of bench terrace: Level bench terrace, slope inward and slope outward bench terrace.

ii) Terrace spacing

Terrace spacing is the vertical distance between two successive bench terraces (Fig.66.2). So, terrace spacing is expressed as the vertical or horizontal distance between two adjacent terraces. The horizontal distance is obtained by dividing vertical interval by land slope (m/m). Terrace spacing is depending on the soil depth, land slope, surface condition, and infiltration capacity of the soil. Terrace spacing is equal to the double of the depth of cut. The depth of cut is influenced by the vertical interval (V.I).

iii) Width of bench terrace

The width of bench terrace is depending on needs. Once the width of bench terrace decided, the depth of cut or spacing of bench terrace may easily be calculated.

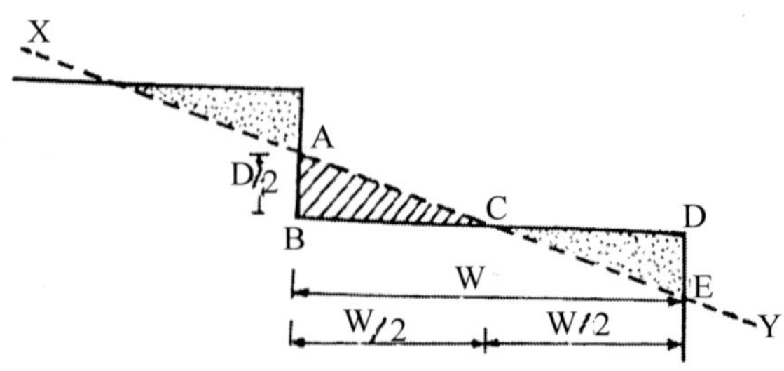

Fig.66.2. Bench terrace cross-section

Let,

V.I = vertical interval between two successive bench terraces

S = land slope, %

W = width of bench terrace

d/2 = depth of cut

D = 2d is the vertical drop between two consecutive terraces

Case 1: When terrace cut is vertical, the depth of terrace may be written as:

$$D = \frac{W.S}{100} \qquad (1)$$

Case 2: When terrace cut or riser slope is 1:1, the vertical drop D = W + D

$$W + D = \frac{100D}{S} \qquad (2)$$

By rearranging the equation, the depth of terrace is given by:

$$D = \frac{W.S}{(100 - S)} \qquad (3)$$

Case 3: When terrace cut or riser slope is 1/2: 1, the vertical drop D = W + D/2

$$W + \frac{D}{2} = \frac{100D}{S} \qquad (4)$$

By rearranging the equation, the depth of terrace can be written as follows:

$$D = \frac{2W.S}{(200 - S)} \qquad (5)$$

iv) Terrace cross-section

The design of bench terrace cross-section includes determining the land slope, dimension of shoulder bund and size of drainage channel.

a) Land slope

The topography of the land is main factor that determined the width of bench terrace. For land slope in the range of 10 to 15%, approximately 8 m width of platform should be constructed. For slope range of 15 to 25%, the width is reduced to about 5 to 8 m and for a steeper slope of 25 to 33%, the width should be further reduced to about 3 m.

b) Shoulder bund

The shoulder bund is constructed with a trapezoidal shape. Bund size is kept nominal in terraces sloping inward whereas in terraces slopping outward and level top, the size of bund should be larger cross-section for holding the runoff water. The height of bund ranges from 15 to 30 cm and bottom width of 75 cm is provided for side slope of 1:1. For bench terraces sloping outward, recommended height, bottom width and top width are 45, 120 and 30 cm respectively.

c) Drainage channel

To drain the excess rainfall from the terrace platform, the drainage channel is constructed either riser wall side or shoulder bund side. The size of drainage channel is depending upon the annual rainfall, soil permeability and peak discharge.

Peak discharge

Peak discharge can be determined by the following steps:

Step1

The area of terrace that has to be drained by the channel is calculated by the following formula:

$$A = \frac{L \times W}{10000} \quad \text{..........(6)}$$

Where,

A = area to be drained, ha

L = length of terrace, m

W = width of terrace, m

Step 2

The peak discharge can be calculated by using any standard method, such as rational method. The design should be based on 10-years recurrence interval.

$$Q = \frac{CIA}{360} \quad \text{..........(7)}$$

Where,

Q = peak discharge, m^3/s

C = runoff coefficient

I = rainfall intensity, mm/h

A = area of terrace, ha

Step 3

Cross-sectional area (A) of the channel can be calculated approximately as the value of Q and V are known.

Step 4

The channel capacity can be calculated using the Manning's formula:

$$Q = A \times \frac{R^{2/3} \times S^{1/2}}{n} \quad ....(8)$$

Where,

Q = design runoff rate
A = cross-sectional area of the channel
V = flow velocity
R = hydraulic radius, (A/P)
P = wetted perimeter
S = slope in fraction (or) channel grade
n = Manning's roughness coefficient (0.004 for design purpose)

v) Terrace length

In general, 400 to 500 m is the maximum distance that a terrace should carry water in one direction.

3. Area lost in bench terracing

Using the following empirical formula the area lost can be calculated.

(i) When batter slope is 1:1, the area lost can be calculated by:

$$A_L = \frac{S + 200}{\frac{200}{S} + \frac{S}{100}} \quad ....(9)$$

(ii) When batter slope is ½:1, the area lost can be given by

$$A_L = \frac{S + 100}{\frac{200}{S} + \frac{S}{100}} \quad ....(10)$$

Where,

A_L = percentage area lost
S = land slope in percent

Example 1: If the vertical interval between two bench terraces is 3.5 m and batter slope is 1:1, calculate the following parameters (i) width (ii) length per hectare (iii) earthwork and (iv) area lost. The bench terrace is proposed to construct in the slope of the hilly areas of 17 per cent.

Solution

Given data: Vertical distance between the bench terraces (D) = 3.5 m, S = 17%

For a batter slope of 1:1

i) Width of bench terrace (rearrange Eqn. 15.22)

$$W = \frac{D(100-S)}{S} = \frac{3.5(100-17)}{17} = 17.088m$$

ii) Length of bench terrace/ha

$$L = \frac{10,000}{W} = \frac{10,000}{17.088} = 585.206m$$

iii) Earthwork per hectare = area of bench terrace × length of bench terrace

$$= \frac{1}{2} \times \frac{W}{2} \times \frac{D}{2} \times L$$

$$= \frac{1}{2} \times \frac{17.088}{2} \times \frac{3.5}{2} \times 585.206 = 4375m^3$$

(iv) Area lost due to terracing

$$A_L = \frac{S+200}{\frac{200}{S}+\frac{S}{100}} = \frac{17+200}{\frac{200}{17}+\frac{17}{100}} = 18.18\%$$

Example 2: Design a 200 m long bench terrace for a land having an average slope of 20%. The soil is clay loam. The terrace channel has a uniform grade of 0.5%. Maximum intensity of rainfall expected during the 10-year recurrence interval is 15 cm/hr and runoff coefficient is 0.6. The width of terrace is assumed as 5 m.

Solution

Given data: L = 200 m, S = 20%, W = 5 m, I = 15 cm/hr, C = 0.6

As the soil type is clay and rainfall intensity is 15 cm/h, a slope inward type terrace is selected. The inner end of the terrace is 30 cm lower than the outer end.

The slope of terrace face (riser) is 1:1 (selected as per condition)

i) Vertical interval between terraces:

$$V.I = D = \frac{W.S}{(100-S)} = \frac{5\times 20}{(100-20)} = 1.25\text{m}$$

ii) Area of terrace that has to be drained by the channel:

$$A = \frac{L\times W}{10000} = \frac{200\times 5}{10000} = 0.1\,\text{ha}$$

iii) Design discharge:

$$Q = \frac{CIA}{360} = \frac{0.6\times 150\times 0.1}{360} = 0.025\text{m}^3/\text{S}$$

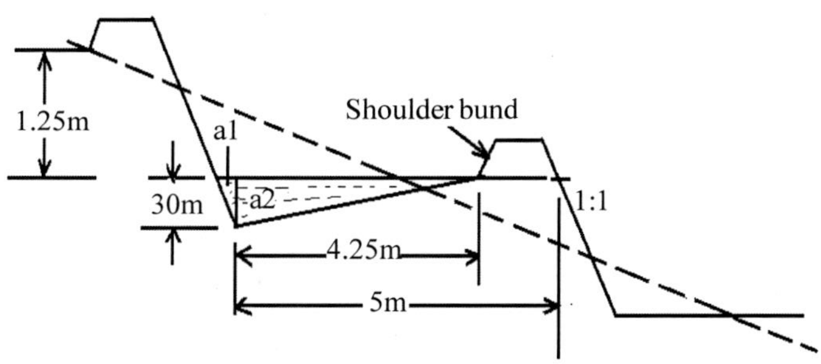

Fig.66.3. Design of bench terrace

iv) Cross section of the shoulder bund (as per practice):

Side slope = 1:1; Height = 30 cm; Top width = 15 cm; Bottom width = 75 cm

Referring to Fig.66.3,

$a_1 = ½ \times 0.3 \times 0.3 = 0.045\ \text{m}^2$

$a_2 = ½ \times 0.3 \times 4.25 = 0.6375\ \text{m}^2$

$a = a_1 + a_2 = 0.045 + 0.6375 = 0.6825\ \text{m}^2$

$$S_1 = \sqrt{0.3^2 + 0.3^2} = 0.42\text{m}$$

$$S_2 = \sqrt{0.3^2 + 4.25^2} = 4.26\text{m}$$

Wetted perimeter, $p = s_1 + s_2 = 0.42 + 4.26 = 4.68$ m

Hydraulic radius, R = a/p = 0.6825/4.68 = 0.1458 m

Mean velocity of flow,

$$V = \frac{R^{2/3} \times S^{1/2}}{n} = \frac{0.1458^{2/3} \times (0.5/100)^{1/2}}{0.04} = 0.4897 \text{ m/s}$$

The velocity is non-erosive.

Carrying capacity, $Q = a \times v \quad = 0.6825 \times 0.4897 = 0.3342 \text{ m}^3/\text{s}$

67

Estimation of Soil Loss by Universal Soil Loss Equation

Aim

- To estimate the soil loss using universal soil loss equation

Universal soil loss equation

Wischmeir and Smith (1962) has been developed a predictive equation for estimating the soil loss taking into account following factors: Climatic characteristics, soil characteristics, crop management and conservation practices. It is called Universal soil-loss equation (USLE).

$A = R \times K \times L \times S \times C \times P$

Where,

A = average soil loss for the given period (tones/hectare/year)

R = rainfall erosivity factor, which is the measurement of the kinetic energy of a specific rain event.

K = soil erodibility factor. It is the rate of soil loss per erosion index unit for a given soil as measured on a unit plot (22.1 m long with 9% slope in continuous clean-tilled fallow)

L = length of slope factor. It is the ratio of soil loss from the field plot under exiting slope length to that from the 22.1 m slope length (unit plot) under identical conditions.

S = steepness of slope factor. It is the ratio of soil loss from the field slope gradient to that from the 9% slope (unit plot) under identical conditions.

C = crop management factor. It is the ratio of soil loss from the area under specified cover and management to that from an identical area in tilled continuous fallow (unit plot).

P = soil conservation practices factor. It is the ratio of soil loss under a support practice (terracing, strip cropping and contouring) to that under staight-row farming up and down the slope.

Rainfall factor (R)

- Rainfall factor is referred to as rainfall erosivity index, which express the vulnerability of rainfall to erode the soil from an unprotected field.
- Soil loss from cultivable land is directly proportional to the product of the two rainfall characteristics such as (i) kinetic energy of the storm (ii) maximum 30-minutes rainfall intensity.
- The product of kinetic energy of the storm and the 30-minute maximum rainfall intensity is termed as EI_{30} index which gives the best estimation of soil loss.
- The highest average rainfall intensity in any 30-minute period during the storm, may be computed from recording type rain gauge charts by knowing the maximum amount of rain which falls in period of 30 minutes interval and then converting in intensity (inch/h).
- This method of estimating the rainfall erosivity is referred as EI_{30} index method.

EI_{30} index method

Kinetic energy can be estimated by the following equation:

$$KE = 916 + 331 \times \log_{10}(I)$$

$$EI_{30} = KE \times I_{30}$$

Where,

KE = kinetic energy (ft. tones/acre inch)

I = intensity of rainfall for 30-minutes period (inch/h)

EI_{30} = rainfall erosivity (ft. tones/acre)

K.E > 25 index method

- This method is based on the concept, that the erosion takes place only at threshold value of rainfall intensity.
- From the experiment, it was obtained that the rainfall intensities less than 25 mm/h are not able to detach the soil particles.
- Thus this method takes care only those rainfall intensities greater than 25 mm/h

Kinetic energy can be estimated by the following equation:

$$KE = 210.3 + 89 \times \log_{10}(I)$$

$$EI_{30} = KE \times I_{30}$$

Where,

KE = kinetic energy (m. tones/hectare cm)
I = intensity of rainfall for 30-minutes period (cm/h)
EI_{30} = rainfall erosivity (m. tones/hactare)

Soil erodibility factor (K)

- Soil erodibility factor is closely related to the various soil properties that susceptible to erode the soil particles.
- Physical properties of the soil such as permeability, infiltration rate, soil texture, size structure, stability, soil depth and organic content are greatly influence the rate of soil erosion.
- The soil erodibility factor is expressed as tones of soil loss per hectare per unit of rainfall erosivity index, for a field slope of 9% and 22 m length.
- The soil erodibility factor can also determined directly for a particular location using a set runoff plots of above specification (9% slope and 22 m length).

The following formula is used for estimation of soil erodibility (K) factor value:

$$K = \frac{A_o}{S \times \Sigma EI}$$

Where,

K = soil erodibility factor
A_0 = observed soil loss
S = slope factor

$\sum$ EI = total rainfall erosivity index

Length and steepness of slope factor (LS)

- The slope length has direct relation with the soil loss. It is approximately equal to the square root of the slope length ("L) for the soils on which runoff rate is not affected by the length of slope.
- In general, steep slope increases the soil erosion by increasing the velocity of runoff water which allows more soil to detachments and transports along with surface flow. Surface detention of water also reduced as slope increases.
- Length and steepness of slope factor is combined together and is termed as topographic factor.
- The values of topographic factor is defined as the ratio of soil loss for a specific steepness and length of slope (9% slope and 22 m length) to the corresponding soil loss from a continuous fallow land.

Smith and Wishchmeir (1962) developed following equation for estimating the topographic factor (LS):

$$LS = \frac{\sqrt{L}}{100}\left(0.76 + 0.53S + 0.076S^2\right)$$

Where,

L = slope length (ft)

S = land slope (%)

Smith and Wishchmeir (1978) again developed following equation for LS factor based on the soil loss data obtained from the crop land on slope ranging from 3 to 18% and length 10 to 100 m:

$$LS = \left[\frac{\lambda}{22.13}\right]^m \left(65.41\sin^2\theta + 4.58\sin\theta + 0.065\right)$$

where,

λ = slope length (m)

θ = slope angle

m = variable depends on the steepness of land slope

Crop management factor (C)

- The crop management factor may be defined as the expected ratio of soil loss from a cropped land under specified condition to the corresponding soil loss from a continuous fallow provided that the soil type, slope and rainfall conditions are identical.
- The factors that affect the crop management factors are type of crops, quality of cover, root growth, water use by growing plants and variation in rainfall distribution within the year.
- The erosion control effectiveness of each crop and cropping practices is evaluated on the basis of five recommended crop stages as under:

Stage 1: It includes summer ploughing and seed bed preparations (rough fallow)
Stage 2: Period from seeding to 1 month thereafter (seed bed)
Stage 3: Time period from 1 to 2 months after seeding (establishment)
Stage 4: Period from 2 to crop harvesting time (growing period)
Stage 5: The period of crop harvest to summer ploughing (residue or stubble)
For estimating the crop management factor, the soil loss from each stages should be calculated separately for a particular crop.

Conservation practice factor (P)

- The conservation practice factor may be defined as the ratio of soil loss for a given practice to that for up and down the slope conservation practices.
- The up and down slope conservation practices consists of mainly contouring, terracing and strip cropping, in which contouring appears most effective practice for medium slope ranges from 2 to 7%.
- The effectiveness of contour tillage to reduce soil loss becomes less as the land slope decreases from medium to zero.
- When the land slope increases from medium to steep slopes, contour row capacity diminishes until at steep slope.
- In strip cropping, the medium strips alternate with grain strips tends to slow down the surface flow and thereby catch of eroded soil from the cultivated strips.
- Terraces in hilly area intercept the surface flow down the slope before it attains an erosive velocity to damage the land.
- From field observations, it has been found that when strip cropping practice is adopted with the terracing practice, becomes more effective in erosion and soil loss control.
- The value of conservation practice factor for contouring, contour strip cropping and terracing presented in Table 67.1

Table 67.1. Value of conservation practice factor for contouring, contour strip cropping and terracing

S.No	Slope (%)	Contouring	Strip cropping	Terracing
1.	1.1-2.0	0.6	0.3	—
2.	2.1-7.0	0.5	0.25	0.1
3.	7.1-12.0	0.6	0.3	0.12
4.	12.1-18.0	0.8	0.4	0.16
5.	18.1-24.0	0.9	0.45	—

Example 1:

Find out rainfall erosivity for the following given rainfall data: Amount of rainfall (inch): 2.0, 1.5, 0.75, 0.4 and average rainfall intensity (inch/h): 0.5, 1.5, 2.5, 3.5. Assume, intensity of rainfall for 30-minutes period is equal to 0.75 inch per hour.

Solution

Rainfall (inch)	**Intensity (inch/h)**	**Kinetic energy (KE) (ft. tines/acre)**	**Total KE**	**$EI_{30} = E \times I_{30}$**
2.0	0.5	$KE = 916+(331 \times \log_{10} 0.5) = 816$	$816 \times 2 = 1632$	$4319 \times 0.75=$
1.5	1.5	$KE = 916+(331 \times \log_{10} 1.5) = 974$	$974 \times 1.5 = 1463$	3239 ft.
0.75	2.5	$KE = 916+(331 \times \log_{10} 2.5) = 1048$	$1048 \times 0.75 = 786$	tones/ acre
0.4	3.5	$KE = 916+(331 \times \log_{10} 3.5) = 1096$	$1096 \times 0.4 = 438$	
			4319	

Example 2:

Calculate the energy of rainfall which has intensity 8.0 cm/h and depth 6.45 cm.

Solution:

Kinetic energy of rainfall (KE)

$= 210.3 + 89 \log_{10}(I)$

$= 210.3 + 89 \log_{10}(8)$

= 290.675 m. tones/ha-cm

Energy/ha

= KE × depth of rainfall

$= 290.675 \times 6.45$

= 1874.85 m. tones/ha

Example 3:

Calculate the energy of rainfall which has intensity 4.5 inch/h and total rainfall depth 1.5 inch using EI_{30} index method. Assume the value of maximum rainfall intensity for 30-minute duration as 0.9 inch/h.

Solution:

I = 4.5 inch/h; Depth of rainfall = 1.5 inch; I_{30} = 0.9 inch/h

Kinetic energy of rainfall (KE)

$= 916 + 331 \log_{10}(I)$

$= 916 + 331 \log_{10}(4.5)$

= 1132.213 ft. tone/acre-inch

Energy/acre (E)

= KE × depth of rainfall

$= 1132.213 \times 1.5$

= 1698.32 ft. tones/acre

$EI_{30} = E \times I_{30}$

$= 1698 \times 0.9$

$= 1528.49$ ft. tones/acre

Example 4:

Calculate the value of soil erodibility factor (K) using following observations: rainfall intensity = 6.5 cm/h, maximum 30-minutes rainfall intensity of the storm = 0.9 cm/h, observed soil loss = 6.5 tones/ha, LS factor = 1.2, and assume depth of rainfall is 4 cm.

Solution:

Energy/hectare

$= KE \times$ depth of rainfall

$= (210.3 + 89 \log_{10} I) \times 4$

$= (210.3 + 89 \log_{10} 6.5 \times 4$

$= 282.65 \times 4 = 1130.6$

Rainfall erosivity index (R)

$$R = \frac{KE \times I_{30}}{100} = \frac{1130.6 \times 0.9}{100} = 10.18 \text{m. tones / ha}$$

Soil erodibility factor (K)

$$K = \frac{A_o}{S \times \sum EI} = \frac{6.5}{1.2 \times 10.18} = 0.532$$

Example 5:

Calculate the value of topographic factor (LS) if the land slope is 7.5% and slope length is 200 ft.

Solution:

$$LS = \frac{\sqrt{L}}{100}\left(0.76 + 0.53S + 0.076S^2\right)$$

$$LS = \frac{\sqrt{200}}{100}\left[0.76 + (0.53 \times 7.5) + \left(0.076 \times 7.5^2\right)\right] = 1.274$$

Example 6:

Calculate the value of topographic factor (LS) if the land slope angle is 20 degree and slope length is 250 m. Assume variable (m) = 0.2

Solution:

$$LS = \left(\frac{\lambda}{22.13}\right)^{m}\left[65.4\sin^2\theta + 4.56\sin\theta + 0.065\right]$$

$$LS = \left(\frac{250}{22.13}\right)^{0.2}\left[65.4\sin^2 20 + 4.56\sin 20 + 0.065\right]$$

$LS = 1.47 \times [(65.4\times0.117) + (4.56 \times 0.342) + 0.065] = 13.64$

Example 7: A specified watershed has following informations: contouring is practiced for conservation of soil, field slope length is 250 ft with average slope as 8%, rainfall fall intensity is 4.5 inch/h, maximum 30-minutes rainfall intensity of the storm is 0.8 inch/h, depth of rainfall is 2.6 inch, observed soil loss is 4.3 tones/acre, crop management factor is 0.3 and the value of conservation practice factor for contouring is 0.6. Estimate annual soil loss using Universal soil loss equation.

Solution:

$L = 250$ ft $S = 8\%$ $I = 4.5$ inch/h $I_{30} = 0.8$ inch/h

depth of rainfall = 2.6 inch $A_o = 4.3$ tones/acre $C = 0.3$

$P = 0.6$

Energy/hectare

$= KE \times$ depth of rainfall

$= (916 + 331 \log_{10} I) \times 2.6$

$= (916 + 331 \log_{10} 4.5 \times 2.6$

$= 1132.213$

Rainfall erosivity index (R)

$$R = \frac{KE \times I_{30}}{100} = \frac{1132.213 \times 0.8}{100} = 9.06 \text{ ft. tones / acre}$$

Soil erodibility factor (K)

$$K = \frac{A_o}{S \times \sum EI} = \frac{4.3}{1.6 \times 9.06} = 0.3$$

Slope length factor (LS)

$$LS = \frac{\sqrt{L}}{100}\left(0.76 + 0.53S + 0.076S^2\right)$$

$$LS = \frac{\sqrt{250}}{100}\left[0.76 + (0.53 \times 8) + (0.076 \times 8^2)\right] = 1.6$$

Universal soil loss equation

$A = R \times K \times L \times S \times C \times P$

$A = 9.06 \times 0.3 \times 1.6 \times 0.3 \times 0.6 = 0.783$ tones/acre/year

Exercise: A land having slope of 9%, length of slope is 120 m and contouring is practiced as soil conservation measures. In this place, rainfall fall intensity is 3.5 cm/h, maximum 30-minutes rainfall intensity of the storm is 0.7 cm/h, depth of rainfall is 3.5 cm, observed soil loss is 2.5 tones/hectare, crop management factor is 0.08 and the value of conservation practice factor for contouring is 0.6. Estimate annual soil loss using Universal soil loss equation.

Bibliography

Amin, M.N., Ahammed, S., Roy K.C. and Hossain, M.A. (2005). Coefficient of friction of pulse grains on various surfaces at different moisture content. International Journal of Food Properties, 8(1), 61-67.

Arora, K.R. (1993). Surveying Volume-1 (3rd ed.). Standard Book House, Delhi, India.

Bansal, R.K. (2005). A Text Book of Fluid Mechanics and Hydraulic Machines. Laxmi Publications, New Delhi, India.

Bohanon, H.R., Rahilly, C.E and Stout, J. (1993). The Greenhouse Climate Control Handbook. Acme Eng. and Manufacturing Corp, Form C7S, Muskogee, Oklahoma, USA.

Bourne, M.C. (1982). Food Texture and Viscosity. New York: Academic Press.

Carson, J.K. (2015). Thermal conductivity measurement and prediction of particulate foods. International Journal of Food Properties, 18(12), 2840-2849.

Chakraverty, A. (1997). Post Harvest Technology of Cereals, Pulses and Oilseeds (3rd ed.). Oxford & IBH Publishing Co. Pvt. Ltd., New Delhi, India.

Chakraverty, A., and Singh, R.P. (2014). Postharvest Technology and Food Process Engineering. CRC Press: New York.

Chandramal, B. (1995). Introduction to Soil and Water Conservation Engineering (Surveying, Irrigation, Drainage and Soil Conservation). Kalyani Publishers, New Delhi, India.

Dai, F., Han, Z., Zhang, F., Zhao, W., Gao, A and Li, X. (2015). Determination testing of seed hardness of staple breeding wheat seed in gansu province of China. Advance Journal of Food Science and Technology, 7(4), 260-265.

Das, G. (2000). Hydrology and Soil Conservation Engineering, Prentice Hall of India Private Ltd., New Delhi, India.

ecourses.vtu.ac.in/nptel/courses/Webcourse.../parshall_flumes.pdf

ecoursesonline.iasri.res.in/mod/page/view.php

http://nptel.iitm.ac.in/courses/IIT-MADRAS/Hydraulics/pdfs/Unit2/2_29.pdf

http://www.campdenbri.co.uk/videos/breadquality-testing-video.php

https://pslc.ws/macrog/dsc.htm

https://www.ametektest.com/learningzone/library/articles/what-is-food-texture-and-how-is-it-measured

Hwang, M.P and Hayakawa, K. (1979) A specific heat calorimeter for foods. Journal of Food Science, 44(2), 435-438.

Iguaz, A., Martin, M.B.S., Arroqui, C., Fernndez, T., Mate, J.I. and Virseda, P. (2003). Thermophysical properties of medium grain rough rice (*LIDO cultivar*) at medium and low temperatures. European Food Research Technology, 217, 224-229.

James, Larry G. (1988). Principles of Farm Irrigation System Design, John Wiley and Sons, Inc., New York: 311-317.

Kanetkar, T.P and Kulkarni, S.V. (2004). Surveying and Levelling Part-1 (Reprint). Pune Vidyarthi Griha Prakashan, Pune, India.

Kumar, P., Coronel, P., Simunovic, J. & Sandeep, K.P. (2008). Thermo-physical and dielectric properties of salsa con queso and its vegetable ingredients at sterilization temperatures. International Journal of Food Properties, 11(1), 112-126.

Lal, R and Datta, A.C. (1999). Agricultural Engineering (Through Worked Examples). Saroj Prakashan, Allahabad, India.

Lan, Y., Fang, Q., Kocher, M. F. & Hanna, M. A. (2000). Thermal properties of tapioca starch. International Journal of Food Properties, 3(1), 105-116.

Michael, A.M. (2008). Irrigation Theory and Practice. Vikas Publishing House Pvt. Ltd. New Delhi.

Moya, M., Aguado, P.J and Ayuga, F. (2013). Mechanical properties of some granular agricultural materials used in silo design. International Agrophysics, 27, 181-193.

Ojha, T.P and Michael, A.M. (2011). Principles of Agricultural Engineering Volume-1 (6th ed.). Jain Brothers, New Delhi, India.

Otten, L., Samaan, G.Y.I Ezeike, G.O.I. (1980). Determination of the specific heat of agricultural materials: Part 1. Continuous adiabatic calorimeter. Canadian Agricultural Engineering, 22, 21-24.

Radha Manohar, K and Igatidnathane, C. (2007). Greenhouse Technology and Management (2nd Ed.), BS Publications, Hyderabad, India.

Rao, M.A., Rizvi S.S.H., Datta, A.K and Ahmed, J., (Eds.). (2014). Engineering Properties of Food (4th Edition). CRC Press, Boca Raton.

Roday, S. (2012). Food Science and Nutrition (2nd ed.). Oxford University Press, New Delhi, India.

Sahay, K.M and Singh, K.K. (1999). Unit Operations of Agricultural Processing. Vikas Publishing House Pvt. Ltd., New Delhi, India.

Sahin S. and Sumnu, S. G. (2006). Physical Properties of Foods. Springer, USA

Schwab, G.O. Fangmeier, D.D. Elliot, W.T and Frevert, R.K. (1993). Soil and Water Conservation Engineering, John Wiley & Sons, New York, USA.

Sharma, R., Sogi, D.S. and Balasubramanian, S. (2012). Aerodynamic characteristics of unshelled and shelled sunflower seeds: significance of moisture and cultivars. International Journal of Food Properties, 15(1), 1-10.

Singh, R.P and Heldman, D.R. (2014). Introduction to Food Engineering (5th Edition). Academic Press

Sitkei, G. (1986). Mechanics of Agricultural Materials. Elsevier Science Publishing Co., Inc., New York., USA.

Skogerboe, G.V., Hyatt, M.L., England, J.D and Johnson, J.R. (1966). Measuring water with parshall flumes. Reports, Paper 83.

Sreenarayanan, V.V and Chattopathyay, P.K. (1986). Specific heat of rice bran. Agriculture Wastes, 16, 217-224.

Subramanya, K.(1994). Engineering Hydrology, Tata McGraw-Hill Education.

Suresh, R. (1993). A Text Book of Soil and Water Conservation Engineering. Standard Publishing Distributors, Delhi, India.

Wechsler, A.E. (1992). The probe method for measurement of thermal conductivity. In: K. D. Magli et al. (eds.), Compendium of Thermophysical Property Measurement Methods. Springer Science Business Media, New York.

Willits, D. H. (1993). Greenhouse Cooling. North Carolina Flower Growers Bulletin 38: 15-18.

www.hitachi-hightech.com/global/products/science/tech/ana/thermal/descriptions/dsc.html

Young, J.H and Whitaker, T.B. (1973). Specific heat of peanuts by differential scanning calorimetry. Transaction of American Society of Agricultural Engineering, 16, 522-524.

Zhang, B., Tamura, M., Berger-Doyle, J., Chen, P. (2008). Comparison of instrumental methods for measuring seed hardness of food-grade soybean. Journal of Texture Studies, 39, 28-39.

Zambrano, M.V., Dutta, B., Mercer, D.G., MacLean, H.L., Touchie, M.F. (2019). Assessment of moisture content measurement methods of dried food products in small-scale operations in developing countries: A review. Trends in Food Science & Technology, 88, 484-496.